GROWING

FAMILY FRUIT AND NUT TREES

GROWING
FAMILY FRUIT AND NUT TREES

by
Marian Van Atta
with Shirley L. Wagner

Illustrations by Shirley L. Wagner

Pineapple Press, Inc.
Sarasota, Florida

Information in this book is true and complete to the best of our knowledge. All recommendations are made without guarantee on the part of the authors and publisher. Inquiries should be addressed to Pineapple Press, Inc. P.O. Box 16008, Southside Station, Sarasota, Florida 34239.

LIBRARY OF CONGRESS CATALOGING-IN PUBLICATION DATA

Van Atta, Marian, 1924 -
Growing family fruit and nut trees / by Marian Van Atta with Shirley L. Wagner.
    p. cm.
Includes bibliographical references and index.
ISBN 1-56164-001-8 : $10.95
    1. Fruit-culture. 2. Fruit trees. 3. Nuts. 4. Nut trees.
5. Fruit-culture—Florida. 6. Fruit trees—Florida. 7. Nuts—Florida.
8. Nut trees—Florida. I. Wagner, Shirley L., 1941- II. Title.
    SB355.V36   1993
    634—dc20
                                                    93 - 7056
                                                         CIP

Design and composition by Millicent Hampton-Shepherd
Printed and bound by BookCrafters, Fredericksburg, Virginia

For families who plant fruit and nut trees for the future

CONTENTS

Acknowledgments

We hope this new book, *Growing Family Fruit and Nut Trees*, will enrich your lives. In writing this *Living-Off-the-Land* book, we have had much help and encouragement from friends and family: rare fruit club members, who shared their tree-planting experiences; garden club members, who tested our recipes; and especially our husbands, Jack Van Atta and Ed Shores, who offered constant encouragement.

Help has come from our neighbors all over Florida, Georgia, North Carolina, Indiana, Wisconsin (especially my sister, Sally Stock, and the staff at the Milwaukee Museum), the Worthington family in Texas, the Van Attas in Arizona, cousins in California, Washington and Oregon, as well as dear friends in Colorado; Oklahoma; Lillooet, British Columbia; Hawaii; and Australia and our family in Norway and Finland.

Irene Gast, my part-time assistant, recruited the help of family members in New York, New Jersey, Texas and California. Melbourne artist Marjorie Burnup contributed diagrams.

Shirley Wagner, my co-author, spent long hours researching, organizing and drafting material, as well as helping to gather recipes and photographs, from Pen Women colleagues, friends and family. Shirley also served as illustrator for the book and deserves many thanks for long hours spent in drawing. Special thanks go to the Dipperys at Cedar Hill Farm, Sandra and Paul, for their contribution.

We thank our publisher, Pineapple Press, Sarasota, Florida, for their encouragement and patience.

Special heartfelt thanks to garden radio and TV communicators: Frank Jasa, Tom MacCubbin, Robert Vincent Sims and Gil Whitton who have had us on their programs.

We wish our readers the joyful experience of growing and using family fruit and nuts.

Marian Van Atta

Introduction

Family Fruit and Nuts, Now!

Have you ever planted a fruit tree? Tasted truly tree-ripened fruit? The experience is a memorable one and well worth the effort of family members to choose, plant and cultivate the tree. During our travels, my husband Jack and I have heard many tales of family trees and gathered many tips for planting and growing trees, as well as for harvesting and preparing the bounty. I will be passing some of these on to you in this gardening book in the hopes of inspiring you to begin a fruit and nut tree minigarden in your own dooryard, patio or balcony area.

My fondest memories have to do with my experience with "family trees." When I was a child in Milwaukee, Wisconsin, my mother planted an Italian prune plum. Every summer it produced hundreds of tasty, ping-pong-ball-sized fruit. Besides eating the plums fresh, we enjoyed mother's German plum kuchen — a family favorite.

Every July our family visited a wonderful place called Archie's Farm in Menomonee Falls, where we children immediately raced for the cherry trees. They were sour pie cherries but, when tree-ripened, were sweet enough to eat fresh. And how we enjoyed those cherry pies as well!

In Florida, we were excited by the tropical fruit which was all new to us. We discovered guavas and mangoes at our farm in Punta Gorda. At our Melbourne Village homestead, we planted a mini citrus grove. On our four children's birthdays, we dedicated special "birthday trees." The most outstanding of these is Billy's Lakeland lime, a cross between a native Key lime and a kumquat. It has survived many winter freezes.

I must not forget to mention our fruitful time in Lillooet, British Columbia, a valley surrounded by the rock faces of mountains which keep it warm. What wonderful apricots we picked there. The pits sprouted into apricot trees which we planted at our own home.

During a very wet summer there, Jack built a large dehydrator, modeled after Cousin Jeannie's in Oregon. It worked so well that many of the townsfolk brought us their fruit to dry. Our dear friends, Alice and Cecil McEwen, who raised ten children and many family trees, shared their home orchard with us. We ate sun-ripened apricots, pears, peaches, plums, cherries and many apple varieties with joy. We dried fruit and sent it traveling south to Wisconsin and Florida in gift packages.

Special B.C. cherries included the Queen Anne variety, with a pinkish white fruit, which we harvested from trees along a cliffside using tall ladders. Grateful for his nursing care, a former patient at Lillooet Hospital invited the staff to share his greatest treasure — black cherries from a tiny homestead on Texas Creek Road. Jack and I sampled them. Delicious!

When we travel to Wisconsin, we always stop at my Cousin Helen's on Orchard Drive in Indianapolis to sample old-timey apple varieties. Then, we drive on to my sister Sally's in Hales Corners, where we savor the flavor of red apples from trees that have never been sprayed. Off to Cousin Wilma's farm at New Berlin next, where we sample more apple varieties. Over the years, we have sent apple trees to our Wisconsin family for wedding and anniversary gifts. In return, we've enjoyed their fruit with special pleasure.

In Sapphire, North Carolina, we've picked red peaches and antique apples from abandoned farms. All were tasty beyond compare.

Visiting lychee fruit groves in Merritt Island and Stuart, Florida, makes me grateful that I am a food and garden writer. Who else would have been invited to sample four kinds of lychees? Each one was perfect and delicious in its own way.

At our present Florida homestead, we enjoy carambolas, persimmons, papayas, guavas, mangoes, Surinam cherries, avocados, loquats and mulberries, as well as grapefruit, oranges, lemons, limes and tangerines. We have been treated to black sapotes (chocolate-pudding fruit) grown by Gene Joyner, Extension Agent for Palm Beach County, Florida, and Lori Smith of Indian Harbour Beach. We are giving our own black sapote plenty of TLC and watching for its first blooms.

Jack and I have found that growing only one or two fruit trees of the same variety cuts down on diseases which may proliferate in large commercial orchards. And by simply using a strong stream of water from our garden hose, we eliminate many insect problems before they start.

So, now it's up to you. Experience the special flavors of tree-ripened, fresh-picked fruit and nuts, grown without poisons.

Plant your own family fruit and nut trees now!

— Marian Van Atta

P.S. Jack says, "Don't forget to keep planting fruit trees. Even if you think you may move or may be too old to see them fruit, plant them anyway — for your children, grandchildren and others who may someday live in the mini-orchards you've created. Start today. Make a better world for everyone tomorrow."

Cherries, figs, peaches and plums are best when picked fully ripe.

Choosing Family Fruit and Nut Trees

WHY AND HOW

WHY

Unless you are over the age of 50 or you live within easy driving distance of an orchard, grove or country produce market, chances are you probably don't know the "real thing" when it comes to fruit and nuts.

Today, produce offered in supermarkets is often gassed, waxed or tightly prepackaged and mixed with aged or spoiling pieces. And it's expensive. Unsavory looking items — gnarly, spotted and misshapen, only vaguely resembling book illustrations of fruit and nuts — often may be found at organic specialty shops. And these, too, are expensive.

Pesticide scares and fruit boycotts further dissuade the average fruit and nut buyer. Yet, the innate deliciousness and nutritional value of fruit and nuts make them desirable for our health and enjoyment. What can be done? Why not take up a small spade for humankind and start by planting a "family tree"?

What is a family tree? Simple. It's a tree which a family selects, plants and works together to bring to harvest, sharing the process and enjoying the fruits and nuts of their labors. Live alone? Share a tree with a friend. Gardening is fun, good exercise and psychologically therapeutic. As you work to bring your tree to harvest, remember that in greenery is the preservation of the world — or at least the lessening of the greenhouse effect.

But you live in an apartment or condo, you say. Even in urban or suburban areas with little or no tillable soil available to you, there is still hope for the family tree. If you have at least one sunny window, a balcony or a small plot of ground, you can grow containerized, semidwarf or dwarf trees grafted with a number of varieties which will bear fruit and nuts and supply family and friends with nutritious, healthful treats — inexpensively. Get permission to plant fruit trees at your church or local school yard.

HOW

But let's not put the harvest before the hoeing. The first step for any grower, no matter what size the garden, is a thorough knowledge of the growing environment and the trees that will do well in it.

Check local resources first. If you live in a suburban or country area, begin by doing a little reconnoitering in your immediate surroundings or neighborhood. Chat with tree and grove owners. What are the predominate varieties of fruit and nuts being grown? What pests, diseases and other problems have growers experienced? Which fruits and nuts are their favorites? What have been their yields? Ask for samples. It will give growers a chance to show off their expertise.

Healthy mature trees in older neighborhoods are good indicators of which trees will be successful in your garden. For the adventurous, a survey of local ethnic neighborhoods may introduce you to fruits and nuts you never knew existed. The quince, for instance, is a favorite in Portuguese gardens and is often combined with tomatoes, onions and garlic to make a tasty dish.

If the area in which you are now living is different in climate from the area in which you grew up, you may not be able to grow your childhood favorites. Most New York state apple varieties, for instance, won't make it in Florida because they require a dormant or chilling period in order to bloom. Florida's climate is just too warm for them. If you live in Florida, you may have to cultivate a new taste more suited to your new environment. The carambola, a star-shaped, applelike fruit, will make a good apple substitute for Florida residents. For those in cooler climates who relish tropical treats, the feijoa, which tastes like pineapple, makes a good choice and can be grown indoors in containers.

The County Extension Service of the U.S. Department of Agriculture will have a wealth of information on gardening specifically designed for your area. Printed pamphlets will give you the straight-skinny on the best fruits and nuts to select. County extension agents, local nurserymen, and university extension representatives are ready and willing to regale you with growing facts till the harvest comes in. And all this valuable information is free.

If you live in an urban area, begin with a visit to your local library or chamber of commerce for listings of local horticultural and gardening groups. Learn where to contact them for added information. Who knows? You may even want to join. (See Appendix 2 for a list of associations and clubs.)

The library will also have lists of catalog companies specializing in fruit and nut trees. Most of these companies supply catalogs free, and many catalogs contain valuable information on how to select and care for trees. Send for a few. They will prove especially useful if there is no nearby nursery or garden center. Usually, trees purchased from catalogs and shipped in a dormant state are pruned back and have their dampened roots wrapped in cardboard or plastic. Directions for planting them are included along with the company's guarantee of fruitfulness. Trees from garden and nursery centers come with bare roots, root ball in burlap, or potted.

The U. S. Government Printing Office is another excellent source for information on gardening and is available to you no matter where you live. Your library should have a listing of their latest pamphlets, or, you can write directly to the printing office. Each government pamphlet is geared to a specific subject, such as pruning, container gardening or freezing. The pamphlets, most of which are free, cover the full scope of gardening from planting to preparation. Don't pass up this ready resource.

Before selecting a tree for your area, make sure that it is cold-hardy for your planting zone, that is, capable of withstanding the average low temperatures in your locale. Sound planting in this respect will save your pocketbook and guard against horticultural disasters. Check with your local county cooperative extension service to determine in which USDA plant hardiness zone your garden is located. To help you select a tree for your zone, most catalog companies and nurseries designate the hardiness of their trees using these USDA hardiness codes. Apples, sour cherries and American plums like the cold and survive temperatures as low as -30 degrees Fahrenheit. They can be grown outdoors as far north as Zone 3. Citrus does not tolerate temperatures much below 30 degrees Fahrenheit and is limited to Zones 9 and 10 when planted in the ground.

The chill period — when and how long the average minimum temperature is apt to occur — and the approximate dates of first and last frosts are vital pieces of information as well. Some fruit and nuts require a long period of dormancy or chilling before they are able to set buds. Such trees will not bear in warm climates.

Gil Whitton, long-time Florida garden communicator in TV, radio and print, suggests, "When planting apples, two trees are needed for cross-pollination and better fruit set. In Florida the best varieties are: Dorsett Golden, Anna and Ein Shemer. Don't make the mistake of planting northern varieties."

In addition, you'll want to research pests and diseases prevalent locally and purchase only resistant varieties.

Apples suitable for growing in the deep South are grafted to special rootstocks. There are several varieties which have been available for nearly 20 years.

 "As American as apple pie" is a tasty simile, but one which would have less basis in fact if it weren't for Johnny Appleseed. Appleseed, a folklore hero, is based on the life of John Chapman, who planted apple seeds along the Ohio River Valley so that pioneer folk moving westward might have something to snack on along the way. Chapman lived from 1774 to 1845 and planted apple seeds for more than 40 years, helping to proliferate apples in the Ohio area. *The New American Desk Encyclopedia* lists him as an "eccentric." Strange value system we've evolved.

Appleseed's tale is just one of many in the lore of fruit and nut trees. Young girls in colonial times sought to peel apples in one continuous peel. When the peel dropped, it was said to fall into the shape of a future lover's initials.

Northern apples, as well as cherries, peaches, pears and plums, which come in hundreds of varieties, are sometimes grafted onto the wrong rootstock. These trees die after only a few years of fruiting.

Realizing the importance of rootstock to successful growing, the U.S. Department of Agriculture is now doing rootstock research at locations around the country. For information on rootstocks suitable for your area, write to the American Pomological Society, 102 Tyson Building, University Park, Pennsylvania 16802.

Because of this on-going study, some nurseries are now listing rootstocks in their catalogs. Check with your local agricultural agent for details. (See Appendix 1, Tree Sources, for a list of mail order sources.)

Innate hardiness and resistance to pests and diseases are largely a function of the rootstock. Most fruit and nut trees are grafted, which means a whip, bud or branch cutting from one tree is bonded or grafted to the rootstock of another. The rootstock will determine tree size — standard, semidwarf (50 percent smaller than standard), dwarf (6 to 12 feet), or miniature (genetic dwarf under 6 feet) — and its resistances. The graft portion will determine the type and quality of blossoms, fruits and nuts.

If you have space for only one small tree and your choice of fruit requires cross-pollination, as many apples do, then you might want to have two varieties which will cross-pollinate grafted onto a single dwarf rootstock. Grafting is an art and is best left to the nurseryman. Nurserymen in the "grafting arts" have produced apple trees with up to five different varieties growing on one trunk. These are true "family" trees. Various pitted fruits and nuts have also been successfully grafted onto one root-stock. Robert Kourik writes of a "fruit-salad tree" grown in Marin County, California, which produces three types of plums, as well as almonds, nectarines and apricots. Generally, however, it is best to select fruits and nuts specifically suited to your climate. Almond, avocado, citrus, fig, nectarine, peach and pecans love warm climates. Apples, cherries, English walnuts on Carpathian stock, pears and plums prefer cooler weather. See the Tree Buyer's Guide in Appendix 7.

For a really excellent encyclopedia section including popular and common fruits and nuts, see Rosalind Creasy's *The Complete Book of Edible Landscaping*. Entries include information on effort required, growing zones, tree characteristics and use, as well as on growing, purchasing, preserving and preparing tips in a very readable format.

Espaliered peach

Choosing the Planting Location

YOUR MICROCLIMATE

Location. Location. Location. As it is in real estate, so it is in gardening. The location which you choose for your tree means everything. In a container, in the middle of a lot, next to a wall — no matter where you choose to locate your tree, you are subjecting it to a specific microclimate. Fortunately, you have some control over microclimates and what you place in their midst. You are the demigod of your domain, whether it be pot or plot.

Terrain

It behooves you, as demigod, to acquaint yourself with your domain — its cold, warm, wet and dry areas and its various altitudes. For container trees, locate your sunniest window or balcony and take into consideration the water sources available to you.

For those of you with at least a dooryard, take a sheet of graph paper and draw your site to scale. This will help you get a better idea of your microclimate. Note the low and high areas in the terrain, water sources, shady areas, existing vegetation and architecture in or immediately adjacent to your site.

On your drawing, include a directional marker and jot down the direction of prevailing winds. If your area is subject to neighboring influences such as heavy traffic or spraying or if it is bothered by pests or diseases, note them. These jottings will help you choose the best location and care for your tree. (See Appendix 3, Planting Record.)

For instance, trees planted on a slope will experience more run-off and will probably require more water and fertilizer to bring to fruition than trees planted in a low area. Low areas, which are traditionally more fertile, form pockets of dampness, are cooler than surrounding areas and are more readily subject to frosts. A tree designated for Zone 8 might do well in a Zone 9 garden if it is planted in this cooler location.

Plantings on the south or east side of a building may enjoy a climate as much as 10 degrees warmer than trees on the west or north side of the same building. Other structures such as eaves or overhangs keep off chilling rains for early blooming trees in cool climates but limit the size of the tree. Mulches are an additional way to warm the local climate.

Borderline trees designated for warmer zones and miniature trees are more likely to fall prey to pests and disease and are easily frosted. If you have experienced such pests or diseases in the past, treat the area before you plant and choose only trees grafted to resistant rootstocks which are designated for your area.

Soil

Dirt is not soil. Dirt you'll find on your five-year-old's hands. Soil contributes to the climate for tree growth by suppling trees with air, water, nutrients and organic matter.

Your job as a gardener is to bring your existing soil to a close approximation of fertile garden loam. Loam is the type of soil found on the forest floor. It is an air-filled, crumbly soil which promotes nutrient availability, good root penetration and good drainage. One reason for these properties is its high content of organic matter — decayed leaves, animal droppings and the microorganisms which process them.

Before you can improve your soil, you have to know its basic characteristics: the type of topsoil, its depth and its pH. You can determine these traits on your own or take samples to your extension service, garden center, or nurseryman for testing. Such tests are often done free or for a nominal charge.

Whichever method you choose, you should begin by taking a soil sample. Take several soil samples from the site and mix them together to form a good composite soil. Samples should be taken from the top 6 inches of soil and enough should be taken to fill a pint container.

After taking the sample, dig down until the color and texture of the soil changes and measure the distance to the surface from the area of color change. This figure gives you the depth of your topsoil. Take a pint sample from a depth of 6 inches of the lower layer, subsoil, as well. Be sure to label these samples.

If you are attempting to determine the soil composition yourself, begin by taking a handful and rolling it in your palm. If it is predominantly sandy, you'll be able to feel the granules. Sandy soils drain quickly, washing away nutrients and leaving trees prone to drought and poor anchorage, particularly if the subsoil is also sandy. To improve the topsoil, you will have to add fertilizer and compost.

If the soil in your palm feels greasy but doesn't stick together, it's silt. Add some sand and compost. But if it is sticky and seems like a candidate for making pots, it is clay. Clay holds water and nutrients but resists root penetration. To unlock its vital nutrients, aerate the soil by adding sand and compost.

These same tests can be run on the subsoil. The subsoil determines the internal soil drainage. For instance, if the subsoil is sandy, a tree planted in a flat area will require more fertilizer and compost than one planted on a slope in clay soil.

Minerals

Proper tree growth requires a balanced combination of 15 minerals or plant food elements, which are naturally present in the soil in different amounts. The elements present in the least concentrations will determine whether the tree flourishes. For example, if your soil is low in nitrogen, your tree will grow until the nitrogen is used up. In this case, you should add nitrogen to the soil, making sure not to add too much, in order not to upset the balance of elements. If this happens, the mineral present in the lowest concentration, and therefore the one which will determine tree growth, will no longer be nitrogen.

Plant food elements will be discussed in more detail in Chapter 4.

pH

Some trees like slightly acid soils, while others like slightly alkaline soils. Acidity and alkalinity are measured on a pH scale where the midpoint, 7.0, the pH of pure water, is the neutral point. A pH above 7.0 is alkaline; a pH below 7.0 is acidic. If you are a do-it-yourselfer, kits for testing pH are available at your garden center. Most trees prefer soil with a pH in the 5.5 to 7.5 range.

Whether you do it yourself or let the experts at the county

extension service do it, the results will indicate whether a change in soil pH is needed. If your soil is extremely acid or alkaline, you should take steps to bring it toward the neutral point. For acidic soils, this means adding dolomite lime; for alkaline soils, adding sulfur. Both dolomite lime and sulfur are slow to act, and achieving the desired effect may take time.

The evaluation of your terrain and soil will provide you with a "profile" of your domain, enabling you to make the best use of existing features and to improve where necessary before introducing your family tree.

Containers

If you plan to grow your tree in a container, you have the advantage of a more manageable, if miniature, site. Whiskey barrels cut in half make excellent tree containers. There are also many ready-made wooden, plastic and clay containers on the market. If you choose, of course, you can build your own.

The containers should be at least 2 cubic feet in capacity with generous drain holes. Place broken, curved pieces of pottery or wire mesh over the holes to prevent soil spillage, and fill the bottom of the container with a 2- to 3-inch layer of gravel. Two parts commercial potting soil and one part organic compost will make a good planting soil, and peat moss will make a good mulch after the tree has been planted.

If you live in a cool climate, you may want to purchase container trays with wheels or build a wheeled platform for your containers. Castors may be purchased at any hardware store. A wheeled platform makes it easier for you to bring your tree indoors easily during cold periods, to position it to receive maximum sunlight, or to move it for cleaning purposes. Miniature citrus trees with their evergreen foliage make attractive house plants throughout the winter months.

Planting Your Tree

ROOTS FOR SUCCESS

Now that you've chosen your tree, taken into account its hardiness, chill requirements and resistances and considered the microclimate in your yard, including its soil characteristics, I am sure you're quite exhausted. Buck up. Breathe deeply and concentrate. You are about to take the single most important step for the health and growth of your tree: planting it. There is no turning back. You must get it right the first time.

Root Preparation

Rootstock determines the size and resistance of grafted trees, and you've chosen your tree to suit the characteristics of your site. Trees may be purchased in containers, with balled roots or with bare roots. For transplanting, the root portion of the tree should out-balance its top. In fact, the smaller the tree top in proportion to the roots, the better your chances of success. Trim back damaged branches to an outward-facing bud and cut off dead wood flush with its branch.

If you're transplanting a tree from a container, wet it thoroughly before attempting to remove it. Potted trees tend to become root-bound. If trees are placed in the ground in a root-bound condition, they will remain that way, and tree growth will be stunted. After you've freed the tree from the container, unwind its encircling roots, cut off any dead roots, and free up the soil with a fork.

Trees with balled roots in burlap should be unwrapped halfway, and the surrounding soil should be loosened. Fill the hole to the level of the burlap, lay back the burlap edges and continue to fill and tamp the soil.

Bare-root trees, available in early spring, require special care to keep their roots from drying out. If the trees cannot be planted immediately, they should be heeled-in until planting time. Heel-

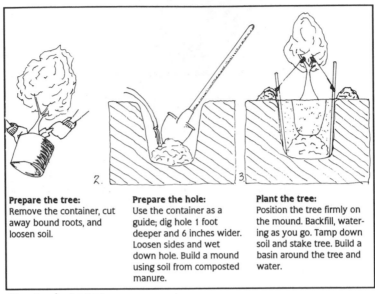

Prepare the tree:
Remove the container, cut away bound roots, and loosen soil.

Prepare the hole:
Use the container as a guide; dig hole 1 foot deeper and 6 inches wider. Loosen sides and wet down hole. Build a mound using soil from composted manure.

Plant the tree:
Position the tree firmly on the mound. Backfill, watering as you go. Tamp down soil and stake tree. Build a basin around the tree and water.

How to plant a tree

ing-in means laying the trees in a trench at a 45-degree angle and covering their roots with soil which is kept moist but not wet. Just before planting bare-root trees, remove dead and broken root tips. Because bare-root trees have the least damaged root system, they have a head start over containerized or ball-root trees. They are, however, more difficult to plant, as you will see.

Site Preparation

Before beginning to dig, recheck your site for architectural obstacles, low-hanging wires and buried cables or pipes. Avoid them. They will interfere with your tree's growth.

If you are planting a tree from a container or with a root ball, dig a hole about a foot deeper than the depth of the roots and twice as wide. Make three piles of soil — sod, topsoil and subsoil. Build a mound of soil in the bottom of the hole using the topsoil. You may want to improve on the topsoil by adding small amounts of mineral fertilizers such as colloidal phosphate or organic matter peat, if your soil is poor. Animal manure does not last as long as colloidal phosphate but provides food for microorganisms which aerate and improve the soil.

There are a number of theories on how best to improve the soil at this point. Some gardeners feel that adding fertilizers will encourage root burn unless you let the hole stand for at least 2 weeks before planting. Some believe in holding back organic additions for the backfill nearer the top of the hole to better feed the hairline roots. Check with your local agricultural extension for the best advice for your area.

Planting

Whatever materials you include in your mound, make it tall enough to raise the tree to the same level as it was while in its container or in the ground at the nursery. The graft portion, or bump, should be above the soil surface and the root system just below it. If you mistakenly bury the grafted portion of a dwarf tree, you could end up with a standard-size tree 60 feet in height instead of one 6 feet tall. This makes for a very crowded patio.

Place the tree in the hole and fill the hole with a mixture of 2 parts subsoil/topsoil to 1 part organic matter, watering it at intervals to ensure good root-soil bonding. Firmly position the tree and tamp the soil down around it so that it rests securely on the mound. Packing the soil tightly in this way eliminates air pockets.

Use the broken ground cover to make a rim or berm around the tree about 5 or 6 inches high and at least 2 feet from the tree trunk. This will form a basin or moat around the mounded tree. Fill the basin with water — up to 5 gallons — and let it drain. Take care not to soak the tree trunk. Wet trunks encourage the dreaded crown rot. See Figure 4 for an example of proper tree planting.

Planting bare-root trees requires two people: one to hold the tree in place and one to shovel. This is no joke. First, prepare the hole and center mound as before. Insert a long stake into the center of the mound. Then place a flat board across the hole near its center. Next, shake out and spread the tree roots over the mound. Line up the old soil mark on the tree trunk with the intersection of the stake and board. This will keep the tree at the proper depth. An alternate method for ensuring that the tree is set in the proper position at the correct depth is to use a planting board with holes cut for stakes and tree trunk. When the tree is

Back in 1922, an Englishman, Richard St. Barbe Baker, founded the Men of the Trees Society in Kenya. Its first members were Kikuya Tribesmen of Kenya, Africa. Today the society has a membership of over 20,000. They have three main tenets:

> Do one good deed a day.
> Plant ten trees each year.
> Take care of trees everywhere.

Since most of us aren't living in the open spaces of Africa, we might moderate these tenets slightly for the dooryard gardening of family trees:

> Do one good deed a day.
> Plant one fruit or nut tree a year.
> Be kind to trees everywhere.

positioned, the second person can fill in the hole with soil as before, alternating watering and tamping.

If you live in a damp climate and the tree you've chosen is susceptible to crown rot, you might want to mound your soil before planting. This means covering the planting site with up to 6 inches of loamy soil before you plant. As this soil settles, it will leave the crown of your tree high and dry, avoiding crown rot.

Two highly experienced fruit growers, Bob and Opal Smith, have planted more than 400 tropical fruit trees by hand, using the "mud-in" method which varies slightly from the standard method. First, they remove sod and weeds in a 6- to 8-foot circle. Then they spread and stomp in 40 pounds of composted cow manure, dig a circular hole the size of the tree container in the center of the circle, and fill it with water. As the water subsides, they add a handful of 20-20-20 slow-release fertilizer to the hole. They prepare the tree by washing the soil from its roots before planting it or "mudding-in," replacing the soil and manure around the tree using one of the methods above.

Timing

Bare-root trees must be planted in the spring after the last frost. Trees balled in burlap or in containers may be planted any time throughout the year. Late spring or early summer is best, however, because this allows the tree to establish itself before blooming the following year.

What about the influences of the moon and planets? It's not lunacy. The phase of the moon coupled with planetary positions does affect the health and vigor of your plantings. Consult your local health food store for booklets containing lunar calendars. Most will list the best times for performing gardening tasks, including planting.

The moon passes through four phases or quarters each month. During the first and second quarters, the moon is growing from a new moon to a full moon, or waxing. During the third and fourth quarters, it is diminishing from a full moon to a new moon, or waning.

During each lunar cycle, the moon simultaneously passes through 11 planetary positions or astrological signs, spending about 2 and one-half days in each sign. Each of these signs has a bearing on the gardening process. For instance, barren signs — Aries, Gemini, Leo, Sagittarius and Aquarius — are good for destroying pests and weeds, pruning and cultivating. Productive signs — Picses, Taurus, Cancer, Libra, Scorpio, and Capricorn — are good for planting and transplanting.

The waxing and waning moon coupled with these planetary signs form the bases for planting guidelines. The best times for planting or transplanting fruit trees is the third quarter of the waning moon in Cancer or Scorpio.

This method is fine if you live at a pace in tune with the tides; otherwise, plant whenever your time and the weather permit. That's what we do.

Nurturing

Newly planted trees should be watered every day for the first 3 or 4 weeks, if they are planted outdoors in the ground. After that, you may taper off, watering 2 or 3 times a week for 1 or 2 weeks, until you are watering only once a week. Fertilize the tree after about 6 weeks. To help keep the young tree free of disease when planted outdoors, keep a radius of 1 or 2 feet from the trunk clear of grass and weeds and do not mulch. Mulches harbor dampness and encourage diseases and insects on trees planted outdoors. You might also want to paint the tree trunk with lead-free exterior latex paint to ward off bugs.

Most, although not all, growers recommend pruning immediately upon planting to bring the top of the tree into balance

Build a berm around the tree to form a basin.

with the roots which have already been pruned in transplanting. Such pruning ensures that the tree top receives adequate nourishment for strong trunk growth in the first year. This is particularly true for unbranched trees or whips. Heading them back to lateral buds will force side branches to grow.

If you have purchased a branched tree, it is probably at least 2 years old and already shaped by the grower. Be sure to ask the age of the tree at the time of purchase. Do not prune back the branches severely unless they are dead, broken or deformed.

You may want to stake the tree to keep it from wobbling around in the hole and losing root contact with the soil. Choose sturdy stakes and sink them at least 2 feet into the ground outside the root ball area of the tree. Fasten the tree to the stakes with nonabrasive materials which are strong enough to secure it without damaging it. Good fasteners may be made from old nylon stockings, rags torn into strips or wire run through old sections of garden hose. The softer the fastener, the more turns around the tree you will have to take to secure it. Use the same type fasteners when securing your bare-root tree to the central stake — the one which you used as a guide in planting your tree.

No matter how or where you've planted your trees, do not allow your young ones to bear more than one or two fruits the first year after planting. Fruit production will sap strength from trunk and root growth and weaken the tree.

Containers

Planting trees in containers is similar to planting them in the ground. Prepare the tree in the same manner. The container should have a 2-cubic-foot capacity and adequate drain holes. You may want to add a 3-inch layer of gravel for drainage. Then fill halfway with a mixture of two parts potting soil and one part compost. Place the tree in the container with the graft or bump even with the container rim. Top up the container with the rest of the soil mixture to within 3 inches of the rim. Tamp down the soil tightly and water thoroughly. If the tree seems insecure, stake it to the sides of the container using the staking method previously outlined. A peat moss mulch works well and is attractive for container-planted trees, which require more moisture than do trees planted outside.

If you are feeling experimental, you might want to try one

of the new nutrient-sac planting methods, such as the Kalliden-dron Method. Using this technique, the tree is planted in a bag which contains a mixture of fertilizer, trace elements and organic matter in a nutrient packet. To prepare the "site" for planting, the grower adds local or purchased garden soil and perforates the bottom of the bag with root/drainage holes.

If you try this method, the next step is to plant the tree in the bag and then plant the bag in a container or in your yard, watering it with up to 10 liters of water. The water will percolate down to the nutrient packet, and the tree will receive a continu-ous flow of nutrients. To keep the tree moist, you must water it at specified intervals.

Keep a record of your plantings listing tree type, rootstock, date of planting, site, conditions, fertilizer, watering, pests/dis-eases and treatment, yield and date, best use and other com-ments. These records may be referred to in future plantings. You'll want to see the Planting Record form in the appendix section.

Planting trees for special occasions such as birthdays or as memorials of persons and events is a practice that should be encouraged. We've included The Golden Harvester Planting Cer-tificate in the appendix to help you record these special dates.

CHAPTER 4

TLC for Your Tree

Now that you've planted your tree correctly, you are off to a good start. You cannot, however, rest on your shovel. Fruit and nut trees require tender loving care to bring forth their bounty. Among the necessary tasks are watering, mulching, feeding, pruning and protecting.

WATERING

Providing adequate water flow to tree roots is a top priority. Deep watering encourages deep root penetration. There are several factors to consider when watering: soil, weather and tree requirements. Sandy soil and hot, dry, windy climates will necessitate more frequent watering, especially if your trees are young, bearing fruit or planted in containers.

Before watering, test for moisture. Treated sticks which indicate soil moisture by changing colors are available at your garden center and are especially good for container-grown trees. Water containerized trees until the water pours from their drain holes — into a drip tray, preferably. Mist the foliage. Remember that dark-colored containers dry faster than light-colored ones and that wood and clay containers dry faster than plastic ones. If your tree loves warmth and moisture, then the ultimate container is a white plastic one.

Outdoors, test for moisture by driving a rod into the soil. In loose soil, the rod will stop abruptly when it encounters dryness. If less than 2 feet of soil moisture shows on the rod, you will need to water. Larger trees demand watering to a depth of 4 feet. You can also take a soil core and examine it for moisture. Special tubes may be purchased for this purpose.

Whichever method of watering you choose, it should direct the water to the tree roots. Watering the tree trunk and branches is not a good idea because it promotes rot.

When you planted your tree, you built a basin or moat

Fruits which ripen after picking — pears, avocados

around it. Enlarge this as the tree grows, keeping the rim just under the tree drip line. Under dry conditions, you may need to build up the rim to increase the capacity of the basin. You can determine whether you need to do this from your tests. As a rule of thumb, in sandy soils, if you fill the basin with an inch of water, it will penetrate 11 inches. In clay soils, an inch of water will penetrate 4 or 5 inches. So, if your soil is sandy and you want to water to a depth of 2 feet, you must build a basin that will hold 2 inches of water at the tree drip line, without flooding the tree trunk.

If rotating lawn sprinklers are used, don't aim them at the tree trunk. This type of sprinkler tends to waste water by spraying it too widely, and if you are drawing your water from a well in a seaside location, it may even promote tree damage by spraying salt-tainted water. Low-volume, ground-level sprinklers, known as emitters, are best. Overhead drip emitters can also be used to ward off frosting. If you choose emitters, the best are mini-sprinklers that give an adequate water cover and are clog-free. Four of these are needed per tree.

Growing trees in arid areas, or in regions that have dry seasons, requires different tactics. Dr. Martin Price at ECHO (Educational Concerns for Hunger Organization) in North Fort Myers, Florida, provides this example of a method used by Kenyan farmers: "The African farmer plants trees in the dry season when the soil is powdery. After planting a tree seedling, he buries an old tin can (with a small hole near the bottom of the side facing the tree). He fills the can with water, which runs out quickly into the dry soil. Then, he fills it again and covers it with a rock to prevent evaporation; thus, he has to water the new fruit trees only once a week. Trees can survive for 2 to 3 weeks without more watering. One farmer established 25,000 trees on his farm and won first place in an African contest for rural tree planting."

MULCHES

Mulches help regulate moisture and warmth, keep away weeds and reduce erosion; many also improve the soil. Unless you live in a very cold climate where trees must be protected from freezing, do not mound up mulch around the tree trunk. Generally, leave a good 5 to 6 inches between the mulch and the tree trunk, since mulches hold moisture and encourage diseases. Never mulch citrus.

Mulching materials include black plastic, tinfoil, brown paper and newspaper strips, wood and bark chips, nut hulls, composted grass, hay or straw, ground cover (periwinkle, clover), tree leaves, gravel, pine needles, peat and sphagnum moss, to name a few.

Mulches for container trees should be pest-free and attractive. If you're growing a subtropical tree in a container, and live in a cool climate, use a black plastic mulch. Tuck it in around the soil, punch holes in it and cover with wood chips. The black plastic will heat up the soil and keep in moisture, cutting down on watering. Peat and sphagnum mosses are also good.

In winter and spring, acclimate your tree by gradually exposing it to temperature change. Build or buy that wheeled platform now if you haven't already done so. It will facilitate moving your plants indoors in winter and outdoors in spring. Outdoors, place your warm-weather tree on the south side of a north-facing wall. Inside, in winter, warm it with artificial fluorescent lights. Lighting kits which produce the red and blue lighting necessary for plant growth are available for as little as $10.

If you're growing a temperate-zone tree in a container in a warm climate, mulch it with aluminum foil and roof it with a

Today, under a program called the Desert to Orchard Project, the Greek government is seeking to improve the society's agricultural and nutritional practices. Their "Kalledendron" method was developed by Professor George Kallistratos, Director of the Laboratory of Nutrition, Ioannina University, Ioannina, Greece. E. Petrakis of the "New Life" Centre for Personal and Social Regeneration in Athens is looking for international cooperation in this and other nutritional projects. The K method is purported to conserve water, fertilizer and work expenditure, as well as avoid pollution.

cheesecloth tent or find an overhang or wall to partially shade it.

FEEDING

Assuming that you've enriched your soil before planting and followed planting instructions closely, your newly planted tree will not need feeding for at least 6 to 8 weeks. What you feed it and when is determined by the type of soil, the climate and the tree.

Soils have major elements, secondary elements and trace elements, as well as organic content. If you've had your soil evaluated, you've probably received a prescription for fertilization. Follow it.

Major elements in all soils are nitrogen (N), phosphorus (P) and potassium (K). The number designation appearing on bags of fertilizer indicates the percentage by weight of these elements. For instance, 10-10-10 indicates that a fertilizer contains 10 percent nitrogen, 10 percent phosphorus and 10 percent potassium. The other 70 percent may contain secondary elements (calcium, magnesium and sulphur) and trace elements (iron, manganese, zinc, copper and molybdenum) in addition to filler material.

Nitrogen is needed for tree growth and foliage, but too much growth during a fruiting period will cut fruit quality and yields. Slow growth and yellowing leaves indicate too little nitrogen, while sprouts and rampant growth indicate too much. The University of Florida Institute of Food and Agricultural Sciences advises us that excessive growth is nonproductive and increases maintenence cost and time spent disposing of yard waste.

Fertilizers are available in dry, liquid or organic forms. Dry, slow-release fertilizers are good for deep-rooting trees planted outside and will not burn the tree. Liquid versions are best for containers. Examples of organic fertilizers are bone meal, blood meal, phosphate rocks and dolomite lime. Compost or composted organic manures improve soil humus content but may contain harmful salts and harbor disease.

Although we have used all types of fertilizer with success at one time or another, we have grown to prefer the liquid form which is more easily handled and transported. We spray our fruit trees from top to bottom with leaf foliar.

Citrus and subtropical fruit and nuts should be fertilized three to four times a year from late winter to late summer with a complete fertilizer. If the tree is 1 to 3 years old, apply 1 to 2 tablespoons of complete fertilizer at four evenly spaced intervals between late winter and late summer. For trees from 3 to 8 years, increase the feeding amounts: use from 1/4 to 1 pound nitrogen fertilizer during the same intervals. For trees over 8 years old, use 1 to 1 1/2 pounds fertilizer applied on the same schedule.

Avocado, fig, loquat, and persimmon need only light feeding. Banana, guava, kiwifruit, lychee, mango, papaya, passion fruit, sapote and citrus are heavy feeders.

Fertilize citrus in Zones 9 and 10 during February, June and October. (Water the tree before applying a fertilizer, so that it won't take up too much fertilizer.) Use a 6-6-6 complete fertilizer applied at a rate of 1/2 pound per inch of trunk diameter at 6 inches above the ground. Spread the fertilizer evenly from a few inches beyond the trunk to the drip line and wash it in thoroughly. Too much nitrogen will make citrus fruits dry and thick-skinned.

Keep a 2- to 6-inch deep mulch of hay, straw, corncobs, pine needles or other available organic material 6 inches from trunk out to tree's dripline. Each spring add a layer of well-rotted manure or 1/2 pound of bloodmeal and 1 pound cottonseed meal for each year of plant's growth. Use half this amount for pears and quinces as too much nitrogen causes fire blight. Now many growers use fish and seaweed leaf sprays. Seaweed added to mulch helps prevent insect and disease problems.

For apples and pears add gypsum and dolomite in small amounts every four or five years.

Bob Whitwood harvests hickory nuts from 75-year-old trees at his Illinois farm and never fertilizes.

Composted manure is somewhat low in nitrogen compared to complete fertilizers. Relatively large amounts must be used to achieve the same growth achieved with complete fertilizers. Bird and rabbit manure are higher in nitrogen than cow manure. For young trees, apply 1/2 pound bird or rabbit manure or 1 pound cow manure in the fall and double the amount until the 7-year point is reached. The same amount must be used thereafter.

If lawn or other ground cover poses a problem, dig holes 12 inches deep and 3 feet apart at the drip line of the tree. Fill them with a mixture of fertilizer and soil.

For containerized trees, use half the recommended amount of fertilizer every 3 weeks from May through June. Pellets or liquid fertilizer are easiest to use.

Custom-mixed fertilizers by Just Fruit, a nursery run by a father-daughter team in northern Florida, have produced excellent results as start-up fertilizers. They make two basic formulas. For acid-loving plants including blueberries and chestnuts, they recommend a formula containing sulfur and iron. For all other fruits, their "sweet" formula contains appropriate amounts of lime. Both mixtures contain gypsum and colloid phosphate to keep the soil loose, fluffy and absorbent. Cottonseed meal, blood meal and bone meal and high-grade container fertilizers which slowly release nutrients ensure that tree roots will not be burned. Real contributors to success are trace elements: zinc, iron, boron and magnesium.

The folks at Just Fruit recommend mixing the fertilizer evenly with the soil in the planting hole. Use 1 1/2 cups fertilizer for trees and 1/2 cup for bushlike fruits. See the Tree Sources section in the appendix for mailing information.

Build your own compost by layering vegetable matter, such as grass cuttings, with improved soil, tamping and watering it. A trash can makes an acceptable container if air holes are cut in the sides at regular intervals to improve air circulation for decomposition purposes. Wire compost cages may also be built or purchased. Into your container, put 6 inches of kitchen waste or grass clippings, 3 inches rabbit manure and 3 inches soil mixture. Moisten, and after 7 days turn it with a shovel. In 3 weeks it will be usable. You can maintain it by adding new peelings or clippings to the compost every 2 weeks. This will promote a gradual decaying process without creating unattractive odors.

PRUNING

Trees are pruned to shape them, to increase fruitfulness and to maintain their health. When you planted your tree, you cut back the roots and headed back the top portion to bring it into balance with the roots. Now, in the first few years of growth, you can shape and train your tree. Basic tree shapes are the pyramid or central leader, the vase or open center shape, and the modified leader shape.

The *pyramid or central leader* method of pruning promotes a dominant central trunk with four evenly spaced branches as the scaffold. The pyramid shape is particularly favored for larger trees such as the pecan or apple tree. The disadvantage of this shape is that it blocks sunlight from interior fruits and leaves unless tree spreaders are used to widen the angles between branches and the trunk. Spreaders are short pieces of wood or metal with a V cut in both ends. These are placed in the tree crotch to increase the angle. Or, limbs may be spread by staking them at a 45-degree angle to a peg sunk in the ground. To avoid

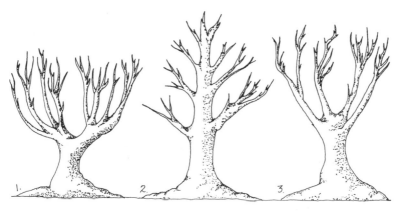

Open Vase Style
Good sunlight penetration. Examples: plums, peaches, apricots.

Central Leader Style
Good tree strength, difficult to harvest. Examples: apples, nuts, miniature trees.

Modified Style
Combines good sunlight penetration and strength. Common.

Pruning and training

damage to the limbs, use wire inserted in a portion of old hose. If the tree seems to be getting too tall, cut the central leader back.

Trees that don't reach great heights, such as the peach or nectarine, are attractively shaped as a *vase or open center tree*. This shape lets in plenty of sunlight. The open center tree has up to five branches, spaced evenly, up to 8 inches apart, which emerge at wide angles to the trunk. All other branches not meeting these criteria should be removed. Cutting back the central leader encourages branch growth.

The *modified central leader* shape is achieved by cutting back the central leader at a height of about 5 feet. Select up to 6 branches along the central leader as scaffold and remove the others. The majority of fruit trees are shaped in this way.

You may also trim your tree as an espalier. Espalier trees are grown in one dimension along a wall, fence or wire. Pruning and spreading them is a bit more complicated than with the other forms. Espaliers may be cordon, horizontal or fan-shaped. To form them, wires are run horizontally along the wall or fence, and bamboo canes are wired to them at the angle chosen for the tree. The tree leader and branches are then tied to the canes with soft string or rag strips. In the case of fan espaliers, the canes are gradually lowered each year until the fan shape is achieved. Apples, pears and peaches are good candidates for this treatment. Branches are selected at the bud stage and unwanted buds are pruned by rubbing them out.

Pruning to increase fruit quality or yield takes into consideration the growth pattern of the tree. Summer pruning increases fruit buds. Winter pruning encourages growth buds. Apples, cherries and pears bear fruit on older wood; therefore, prune them in the summer to keep new and old wood in balance. Plums grow on new wood. Prune them in winter to encourage growth. Thin fruit by removing small or diseased fruit and leaving one or two fruits per spur to increase their size and quality. If a tree is a tip-bearer (one which bears fruit on new wood at the end of the branch, such as a plum), do not prune the tree unless the branches are crowded. Pruning back tip-bearing trees cuts yield.

Maintain the tree by pruning dead and damaged limbs and branches. Cut limbs almost flush with the trunk and seal the wound with pruning sealant. Do not leave ragged, unsealed stubs, because they will attract disease. To head a branch back, cut it diagonally, just above a bud going in the desired direction. Prune suckers, which grow out of the trunk beneath the graft, and also prune water sprouts, which tend to grow vertically and, in citrus, bear large thorns. Sprouts can take over your tree. Also, cut away crowded or distorted limbs and branches.

If the tree has been damaged by a frost, do not prune it immediately. Pruning will stimulate new growth which may be damaged by a late frost. Wait until spring before pruning damaged wood.

PROTECTING

You are the guardian of your tree and must protect it against an onslaught of diseases, insects and animals. The method you choose will go a long way toward protecting yourself and your family from an unsafe or unsavory harvest.

To spray or not to spray: that is the question. We prefer to take advantage of chlorinated city water and insecticidal soap sprays to protect our trees. Many folks who grow fruit trees feel that they will not get edible, insect-free fruit unless they use poisonous sprays.

A growing number of nurseries are at work perfecting trees that require little or no spraying. Stark Brothers Nurseries promotes the Janafree, Liberty and Prima apple trees as fruits which do not require fungicide sprays. They do, however, recommend an insecticide application just after bloom. Miller Nurseries of Canadaigua, New York, lists an apple named Freedom as its most hardy specimen. Freedom has been grown with zero disease for 24 years without the use of controlling sprays. It has resisted apple scab, powdery mildew, cedar apple rust and fire blight. See our list of tree sources in the appendix for the addresses of fruit tree nurseries that provide fruit trees resistant to disease.

If you've already selected your tree, you can help it resist disease by improving the soil and planting it properly. Watering and fertilizing it with organic fertilizer will strengthen it and diminish the likelihood of diseases such as crown rot, crown gall and silverleaf. It will also ward off nematodes.

A nematode is a parasitic worm which lives in the soil and attacks the roots of plants and trees, wilting the leaves. Improv-

There are ways to encourage bumper crops. If your tree is a reluctant bloomer, it can sometimes be shocked into producing by a good whack to the trunk. So reports Robert Vincent Sims of Garden Rebel Radio, a gardening call-in show aired over WWNZ/740 AM, a station serving Central Florida. When a caller asked about encouragement for her frost-damaged avocado, Sims advised her to give the tree a healthy nudge with the bumper of her pickup truck to stimulate its circulation. (Don't let your neighbors see you do this.)

ing the soil with humus is the most common remedy. Planting near a wall or foundation will also deter nematodes.

If you have a recurring problem with nematodes in your soil, you might want to try soil sterilization prior to planting. First rake off the soil. Then plow it with a tiller or shovel. Give it a good soaking so that the moisture will help conduct heat deep into the ground. Then cover the area with clear plastic. Cover the edges with earth so that the plastic won't blow away. Leave the plastic on at least 4 weeks. Temperatures of 120 degrees kill the nasty nematodes!

For fungal leaf disease, remove affected leaves and spray with a solution of 1/2 pound sulfur to 10 gallons water or use a commercial copper-containing fungicide. Bacterial diseases such as fire blight and canker require amputation. Prune away affected branches and heal wounds with pruning sealant. Sterilize your clippers and other tools in a solution of 50 percent bleach. Then spray summer foliage with streptomycin spray.

An ounce of prevention is worth a pound of cure. Spray dormant oil just before buds pop to get the jump on aphids, aphid eggs, mealy bugs, red mites, pear psylla, scale, thrips and white flies. Dormant oil kills by asphyxiation, so give the tree a good coating. This method is good for almost all fruit and nut trees. You can mix your own dormant oil by combining 2 quarts light motor oil with 1/2 quart liquid detergent. Then dilute this mixture with 20 parts water. Or you can purchase a ready-made variety.

Pyrethum, rotenone and ryania insecticides are good for getting rid of soft-bodied insects and are relatively safe since they are nonpersistent or nonsystemic insecticides which do not accumulate in body tissue. Rotenone is also effective for pear psylla, and ryania will terminate moth larvae. Nicotine (tobacco) sprays are good for aphids. A steady stream of water through a hose, or soap and water through a spray attachment, performed regularly, will stop many insect problems before they start.

Natural biological repellants are garlic, tansy, rosemary, rue, beebalm, marigolds and petunias. Plant them in borders or as ground covers, or plant them in pots and place them in juxtaposition to trees in containers. These plants have antiseptic properties and repel aphids, ants and beetles and, in the case of marigolds, nip out nematodes within the drip-line radius of your tree.

Or, you might want to purchase bugs — the beneficial kind — from companies dealing in biological controls. Ladybugs terminate aphids, mealybugs, red spider mites and scale. Larvae of trichogramma wasps consume moth and butterfly eggs, while lacewing larvae take care of aphids, leaf hoppers, mites, thrips and caterpillars. Praying mantises dine on caterpillars also.

Before you purchase these helpful insects, you might want to try attracting them by planting anise, carrots, coriander, dill, fennel, parsley and parsnips. You can also attract bug-eating birds by offering them water and food. Of course, birds aren't very discriminating and may eat beneficial bugs, as well as peck at your fruit.

Microbial agents such as bacillus popilliae and bacillus thuringiensis are good for eliminating grubs and larvae of beetles and moths and may be purchased at most hardware or garden supply stores.

Attracting and trapping insects is another mode of operation. Insect sex hormones, called pheromones, draw lusting insects to sticky death traps. In one case, croquet balls painted red, coated with a sticky glue such as *Stickem* or *Tanglefoot* and hung six to a tree, were reported to reduce insect damage in apples by up to 90 percent. Other sticky innovations include a 9-1 molasses solution in a small container to attract moths.

Fruit flies are quite taken by a solution of fruit vinegar and water in a large-mouth jar. Snails and slugs will creep into strategically placed piles of dampened boards and may be handled at your discretion.

If you can't kill pests, at least make it unpleasant for them. Cinders, bone meal and other rough borders irritate snails, slugs and ants. Of course, you could sink a saucer of beer for snails and slugs, and after they're quite inebriated, do your will upon them.

To discourage burrowers such as rabbits, gophers or moles, place no-kill traps in their tunnels. Then, if you trap an animal, call on the services of the humane society. A tree guard of 1/2-inch wire cloth will protect tree bark and will also protect tree roots if sunk into the ground around the root ball at planting time. Painting the tree trunk with white latex paint likewise acts as a guard.

To scare away birds, hang a fluttery scarecrow, tin strips or tin pie pans on your tree. Inflatable snakes and owls will give birds a fright as well. Or, you can try protecting the tree by looping it with black thread, or draping it with cheesecloth or 3/4-inch nylon netting. Netting will let in light, air and water and is nearly 100 percent effective.

These are just a few of the ways to take out your gardening frustrations. The possibilities are myriad.

Further growing tips are summarized in the chart, Tips for the Organic Grower, courtesy of Gardener's Supply Company. See the appendix section for details.

CHAPTER 5

Harvesting and Distributing Your Homegrown Goodies

Reap what you have sown — and watered, fertilized, mulched and pruned. Your tree's yield, fruit quality and fruit size are directly proportionate to the care you've showered on it.

Generally speaking, the smaller the tree, the more productive it is. As grower, you perform the same tasks no matter what size the tree — standard, semidwarf, dwarf or genetic dwarf. The smaller the tree, the less water, fertilizer and output in dollars you will have to contribute. The time and effort these tasks take is also reduced. Smaller trees produce more fruiting buds per inch of branch than do standard trees. Their yields are greater per given area. And with the latest strains of genetic dwarf trees, the quality and size of the fruits are standard. So small is better — particularly for the dooryard or container grower.

Genetic dwarf trees grow about 6 to 9 feet tall and 6 to 8 feet wide. They are often called miniature trees to distinguish them from dwarf trees which are grafted onto dwarfing rootstock. Genetic dwarf trees have a number of factors in their favor. They have twice as many fruiting buds for branch length as a rootstock dwarf tree, which means lots of attractive flowers. And with fewer leafing buds, you'll find the fruit-to-leaf ratio is high. Sunlight penetration is therefore quite good.

Genetic dwarf trees bear early and their fruit is usually standard size. (Do not allow trees under 3 years old to bear fruit. It will weaken their structure and growth.) You will need no ladders, acrobatic training or stretching exercises to harvest small trees. In some cases, you will actually have to stoop to harvest. There are so many advantages to these little ones that you might actually want to embrace them.

These miniature trees may be trained to the central leader shape because sunlight penetration is not a problem for them.

They will produce more fruit if planted in the ground than in a container, although container-grown trees, given the proper care, will produce enough fruit and nuts for an average family. The affinity of Meyer lemons for containers is well known, and they will produce bumper crops year-round. When a container-grown tree begins to flag, it needs repotting and root pruning to renew its vigor.

The lifespan of genetic dwarf trees can be quite good. One California peach tree, standing 4 feet tall and 8 feet wide, is reported to have produced 300 fruit in one season at 20 years of age.

HARVESTING

Fruits are either climacteric or nonclimacteric, which means they will or will not ripen after picking. Fruits that will ripen after picking include apples, avocados and pears. Fruits that are non-climacteric and are best picked when fully ripe are cherries, figs, nectarines, peaches, plums and most citrus. Most commercially grown fruit is picked in the hard or preripened stage because it will not bruise as easily as ripened fruit in shipping. Fruit may be cosmetically or artificially ripened and waxed or gassed to enhance its beauty. This is why the taste of "real" fruit and nuts just isn't there in many supermarket varieties.

How do you know when a fruit is ripe? Consult your planting guide for harvesting times, which will vary according to the zone, chill factor and microclimate of your site. Citrus ripens anywhere from October to April, depending on its variety and whether you've planted it in Florida, Arizona or California. Generally, however, fruit is ripe when it is vividly colored; when it is firm but not hard; when it is loosely attached to the tree; or, when it has fallen from the tree. Nuts are ready when their outer husks burst open and when they can be easily shaken from their branches.

Harvest time depends on what you're going to do with your bounty after you've harvested it. If you intend to eat it right away, freeze or juice it, leave it on the tree until it is fully ripe. This gives it a chance to accumulate sugar and juices, as well as reach its most radiant aspect. If it is climacteric and you plan to store it, pick it before it ripens.

Chief harvesting methods for the home grower are hand-in-glove, pruning shears, cane, net, or a combination of pole, net, and clippers.

Of the three blessed fruits held in reverence by the Chinese, only the peach has become an important part of the American diet. For the Chinese, it signified longevity. In the U.S. it has come to mean a Southern beauty — a "Georgia peach." The other blessed fruits are the citron or Hand of Buddha, which symbolizes happiness, and the pomegranate, which symbolizes fecundity.

Citrus is ripened when it is in full color, firm and pulls easily from its stem. The best varieties for containers are: Meyer lemon; Mexican, Key or West Indian lime; Temple orange; or kumquat. Whether your tree is planted in the ground or in a container, fruit may be stored right on the tree until you are ready to use it.

Our Valencia orange provides us with fresh orange juice for breakfast from January to May. Limes and lemons fruit throughout the year. Half a Ponderosa lemon will make a fine lemon meringue pie. And several Key limes will add up to a regional treat — Key lime pie. Lemon and lime will also put a little extra zip into your Perrier or Tanqueray. Grapefruit does not store quite as well on the tree, but cooled with overnight dew, it makes a refreshing morning pick-me-up. When fruit skin becomes puffy and soft, it is overripe. This is the time to strip the tree to make way for new blossoms. The same holds true for most citrus.

Cherries are sweet or sour. Sweet cherries require cross-pollinators and are best completely tree-ripened for immediate consumption as fruit or juice. Or, you may temporarily store them in a plastic bag in the refrigerator. Pick them by hand with an upward twisting motion.

Sour cherries grow in bunches and are good for freezing, canning, jamming and piemaking. Cut these off with pruning scissors to avoid breaking tree spurs. Fruit grows on year-old wood, so prune sour cherries in spring to encourage year-old wood the following season.

Peaches and nectarines are nonclimacteric. They should ripen on the tree. They achieve nearly 50 percent of their growth just before ripening. When they give a little around the stalk, bear a pink blush on yellow and show no green, they are generally ready to pick. Lift and twist the fruit to separate it from the stalk.

Peaches bruise easily and are not good travelers. Their shelf-

life is also extremely short. Save them in a cool room or shed on soft material. They may be eaten fresh, frozen, canned or used in pies or preserves.

Plums are ripe when they are firm and have achieved their true colors — purple, red or gold. They may be stored for a few weeks in a cool room, but they are best eaten fresh from the tree. They may be canned, jammed or jellied. If allowed to overripen slightly, prune plums are good for drying in the sun, an oven or in a dehydrator.

Slightly overripe apricots are also good candidates for drying. When they are well-colored, pick them by hand, retaining the stalk, for drying. Avoid pruning the trees. Fruit may be canned if picked before ripening. And apricots will keep about 3 weeks if individually wrapped in newspaper and placed in a cool well-ventilated room.

Pears need cross pollination. They become mushy if tree-ripened. Those that ripen in August or September may be harvested by cutting their stalks while the fruit is still hard. Later varieties may be plucked by the lift-and-twist method. Pears do not travel well and are best eaten after ripening at room temperature for about 3 days. Store them individually on a slatted shelf in a cool shed. They may be canned or preserved also.

Apples, like most other fruits, are best when tree-ripened; however, they are climacteric and will ripen off the tree and store well. They are ripe when they can be lifted and easily twisted from the tree, retaining their stem. A telescoping pole with a net on it may be used to retrieve high-growing apples by knocking them into the net with the net ring. Eat the early fall apples and store or can the later varieties. Sweat apples in a ventilated room for 2 or 3 days before storing them, to get rid of ethylene gases which encourage rotting in closer quarters. You may store apples by individually wrapping them in newspaper or waxed paper and placing them in well-ventilated boxes or a well-ventilated room. Dark, humid, cool places, such as root cellars, are also good.

Figs are ripe when they are soft to the touch and bent at the neck. Beware green figs. Wear protective clothing when picking figs to shield yourself from irritating sap. Figs will refrigerate for a few days for fresh eating. But if you plan to dry them, wait until they drop from your tree or bush. Then the sugar will be at its peak. Peel them only if they are a large, skinned variety;

otherwise, simply quarter or halve them for drying in the sun or in an oven. Dried figs last up to 8 months.

Test avocados for "pickability" by cutting one in half. If the stone has a paper-thin brown covering, it is ripe. Depending on its variety, avocado skin will ripen from green to black or from green to pinkish yellow. When partially colored, avocados may be picked by hand or with an extension claw or rimmed cloth bag. They will ripen at room temperature in 2 to 3 days or keep in the refrigerator for several weeks. Cut them in half, remove the pit and fill their wells with lime juice and salt and pepper for a tasty appetizer.

Nuts are ripe when their husks open. They may be thrashed to the ground or into nets by beating the tree branches with a cane pole or shaking the tree. Pecans in shells will keep a month or so in the refrigerator and forever in the freezer. Shuck and wash walnuts before drying them. Drying takes about 3 weeks. Pick hazelnuts or filberts just as soon as you can twist the husk around them. Dry them in the sun and store them in a cool place.

DISTRIBUTING

If you've produced more than you can eat or store, exchange your gardening hat for a business one. Live on a busy street? Set up a roadside stand in your front yard and let the kids run it under your guidance. It's a great learning experience — a Junior Achievement kind of project. Proceeds can be used for family mad money or tucked away for college or car. You won't have to have a license for a small home stand in most cases, but it is best to check with your extension agent. You might even want to advertise in your local barter or sales newspaper. Community and club newsletters offer additional possibilities.

If you have a bumper crop, load up a pickup or station wagon and drive to a point near a supermarket where you can lower your tailgate and set up shop, offering tree-ripened produce at low prices. Or, you might want to rent a booth at a flea market or a farmer's market. Depending on where you live, you may need a vendor's license to do this.

Try wholesaling. If you've grown organic produce, the local natural food market may be interested in purchasing from you. Mom and pop groceries are another possibility. Food co-ops may take your fruit and nuts on consignment or in trade. If your crop is large enough for them to take notice, processing plants may be interested also.

Until recently, restrictions on fruit disbursement in Florida were a bit more stringent than in other states because of a citrus canker epidemic. Under Florida regulations, if you were a home owner who was giving away or donating citrus fruit, first you had to have your tree inspected for canker. If your tree was declared canker free, you could give your fruit away locally, but you had to wash it first. If you planned to ship it out of state, however, you had to have it dipped by the experts and pay a fee. These regulations are still in effect in a few counties. The state has set up a citrus canker hotline to field questions on this subject: 1-800-282-7161.

The California white fly is another regional pest which has required legislation. In recent years, growers have been afflicted with recurring infestations. In some cases, damage was so extensive that the law required the destruction of fruit to avoid spreading of the larvae.

Other possibilities for bumper crops? Make festive gift baskets for sale or to give to family or friends. Join with others and start a fruit festival. Or, donate the fruits and nuts to churches, charity kitchens, old folks' homes or other worthy causes.

People who live in condos or apartments are particularly appreciative of fruit baskets set out in their lobbies with a "take some" sign affixed. Three bushels of grapefruit were quickly snatched from containers in one such giveaway.

If you can't give it away, turn it into your compost heap and feed it back into the system. Either way, "you've done *good!*"

Enjoying Your Bounty

FRESH, COOKED, JUICED, FROZEN, DRIED OR CANNED

Nothing is better than biting into a crunchy red apple on a crisp fall day, or nipping a cool, juicy peach when it is 90 degrees in the shade — especially if you've just plucked it from your own tree. Fruits and nuts may be enjoyed in a number of ways: as snacks, in salads, as garnishes and flavorings for main dishes, as desserts and as juices.

Fresh

As appetizers and in salads, avocados are delicious halved, filled with Italian dressing and marinated overnight. Mash and combine with lime juice, minced onion, garlic salt and chopped red pepper to create guacamole dip. Avocados are also exciting when sliced thin and served with slices of ripe tomato and spritzed with lemon juice.

Fruit salads are great at breakfast or lunch or as dinner accompaniments. Pears, oranges and grapefruit, mixed together and sprinkled with sugar, make a cooling dish. Vanilla yogurt makes a creamy dressing for peaches, apples and plums. Top with nut pieces.

Fruits and nuts wake up traditional cooking methods. Poultry may be stuffed with apples and walnuts and roasted, or baked with dried apricots and prunes to make savory main dishes. Fish also reaches new flavor heights when baked in parchment with subtropical fruits or grilled with almonds and lime juice. Set off your main dish with garnishes of lemons and oranges or slices of apples and pears.

Fruit and nuts rank high as desserts. Slice peaches into manageable bite sizes, add sugar, stir and refrigerate. They'll make their own syrup and are also good served with cream. Fruit sherbets and ice creams are other delectable possibilities.

Instant Sherbet

2 cups tangelo, tangerine or
    orange juice

2 cups apple cider

Combine the juices and freeze for 20 minutes. Voila! A cooling dessert. If you allow the sherbet to melt slightly, it makes a wonderful slushy. Finish it up at once, because it loses its taste if refrozen.

Who has not hungered for fresh fruit shortcake at one time or another? If you buy individual shortcakes from the supermarket, fruit shortcake is a snap.

Dried

Mix a trail snack from dried nuts and fruit. Ripe nuts should be shelled and distributed over a flat tray and receive full sun for 3 to 5 days. Stir them at least once a day, cover them at night to keep out dampness, and bring them inside if it rains. Oven driers, or racks, may be purchased from gardening suppliers. These fit the standard-size oven and offer three or four parallel drying beds in a wooden frame. Nuts will dry in 3 days in a gas oven with only the pilot light burning or at the lowest setting on your electric oven. Leave the door open just a crack.

Plums, apricots, figs and apples are the fruits most often dried. When they are at the peak of ripeness, peel and halve or quarter them. (Small figs will not require peeling.) Dry them as above for 4 days in the sun, 5 days in the oven, or up to 12 days in a dehydrator. A dehydrator is anything that extracts moisture. A solar, microwave, electric or gas oven will do. When traveling in sunny weather, we have sometimes fashioned a dehyrator on

 The fig, which is an important commercial fruit in California, has a long and revered history. It fed the Israelites in captivity and is a symbol of peace and abundance. In the garden of Bethany, Christ desired figs. Mohammed swore upon the fig tree, and Moslems consider it the tree of heaven, life and knowledge. Its sap is a purgative; cooked in milk, it is an ulcer cure; boiled in barley, a cure for lung infections; mixed with bacon drippings, the bane of mad dogs. In European tradition, the fig fell from holy to profane. To say that you didn't give "a fig" was to show your contempt.

the shelf in the back window of our car by lining it with brown paper and placing our fruit out to sun dry. This is especially effective if you leave the car parked in the sun with the doors closed for any length of time.

Fruit should be sticky and pliable, but not wet. It will keep for 6 to 8 months. You can rehydrate dried fruit by stewing it in water for a breakfast treat.

Cooking

Peel apples halfway and hollow out their cores before stuffing them with a little butter, raisins and honey or brown sugar. Bake until they are tender.

For grapefruit, broiling is best. Cut a room-temperature grapefruit in half and core without piercing the skin. Add a dab of butter to the cavities and sprinkle with cinnamon-sugar to suit your personal taste. Broil from 8 to 10 minutes until sugar is brown and bubbling.

By all means, "let them eat cake." Here are two family favorites to try.

Apple Snack Cake

3/4 cup vegetable oil	1 teaspoon baking soda
2 eggs	1 teaspoon baking powder
2 cups sugar (less will do)	1 teaspoon cinnamon
2 and 1/2 cups all purpose flour	3 cups chopped apples
(substitute 1/2 cup wheat germ if desired)	1 cup chopped pecans

Combine oil, eggs and sugar in a large mixing bowl; beat at medium speed until well mixed. Combine flour, soda, baking powder and cinnamon. Combine the two mixtures. Stir in apples and nuts. Spread batter into a greased 13x9x2-inch baking pan. Bake at 350 degrees for 55 to 60 minutes or until the cake tests done. Cut into bars. (Carambolas may be substituted for apples.)

Here's another finger-licking recipe for fruit cake.

Old Milwaukee German Fruit Kuchen

This recipe can be made with any type of fruit, but it is especially good with Italian prune plums in the North or Java plums in Florida and the tropical islands.

Oil a 9x13 inch pan or its equivalent.

Cake:
2 cups flour
3 teaspoons baking powder
1/2 cup sugar
pinch of salt (may omit)

2 tablespoons butter or margarine
1 egg broken into water to make 1 cup fluid

Topping:
sliced fruit
1/2 cup sugar
2 tablespoons flour

2 tablespoons soft butter
1 teaspoon cinnamon

Mix the first five ingredients with a pastry blender.
Beat egg and water. Add to flour mixture and blend well. Spread batter out evenly in pan.

Put a layer of sliced fruit on top of the cake. Cover with topping mixture of sugar, cinnamon, flour and butter, blended together.

Bake at 375 degrees for 40 to 45 minutes. The cake should be lightly browned. A knife inserted in the center should come out clean.

Juicing

Juices are nutritious thirst-quenchers for breakfast, lunch or dinner. They make refreshing blended drinks and cocktails. Ripe citrus may be juiced by hand or with an electric juicer. Some citrus varieties hold more juice than others and some have more seeds and pulp. Choose the juiciest and sweetest for making fruit juice. Valencia oranges and tangelos fit the bill. Orange juice mixed in thirds with Bordeaux wine and lemon-lime soda and garnished with slices of orange makes an excellent sangria. All orange juice may be drunk without diluting or sweetening it. Grapefruit juice may require a tablespoon of sugar to improve its taste. And lemons and limeades will require both dilution and sugar.

For apple, cherry, and prune juice select tree-ripened fruit. Stem, wash, drain and seed or pit it. Cut the fruit into pieces, cover it with water, and boil it until it is tender. Strain it through a cloth sack. Or use a steam juicer, a three-tiered affair which includes a strainer and cover. You may need to strain twice for clear juice. Firm or hard fruit, which has been cooked, may be liquefied using a blender, if water is added. An apple plus 1 cup of water or another juice will provide 1 1/2 cups of apple-flavored juice. Juice extractors will perform the best job but they are expensive.

In addition to delicious beverages, juices are the starting point for many other drinks, jellies, desserts and flavorings. Fermented, they form ciders, wine and brandies. If you head the household, you might want to investigate these procedures. Try one of a number of books on the subject.

Juice is also the basis for jelly, lemon and lime tarts and pies. Frozen sherbet and sorbet, a mushy, soft-cream treat, are additional recipes which call for nuts as well as juices. And you could try a cool, creamy summer soup, flavored with fruit juice and fruit pieces, such as the following.

Cold Creamy Cherry Soup

1 (1 lb.) can unsweetened pitted sour cherries	1 tablespoon sugar
	2 tablespoons lemon juice
1 1/2 teaspoons cornstarch	1 cup sour cream or plain low-
1/2 cup cold water	fat yogurt

Drain the cherries. Place the juice in a saucepan. Blend cornstarch and water and blend into the juice. Bring to boiling and boil 5 minutes, stirring constantly. Add sugar and lemon juice. Remove from the heat and chill. When the liquid has cooled, blend in the sour cream or yogurt and add the drained cherries. Chill and serve in chilled cups. Yield is about 3 cups.

Freezing

If you can't make use of all the nuts and fruits you've produced, try freezing them. Pecans may be frozen in their shells. Store them in coffee cans with tight-fitting plastic lids. Juices may be frozen in glass or in plastic bottles and jars with screw-on caps. If you do this, make sure you leave a headroom of at least 1 1/2 inches because freezing will expand liquids. Don't freeze juice in containers larger than quart size. They will be unwieldy and may go bad after they've thawed. Fruit juices frozen in freezer trays with cube dividers in place make good ice cubes for drinks and are a handy way to store juices used in cooking, such as lemon and lime juice.

Remember that what comes out of your freezer is only as good as what goes in. Use fruit at its peak ripeness. Wash it, peel it if necessary, trim it and remove the pits or seeds. If is it large, slice it up. Small fruit, such as pitted cherries, can be frozen whole, but most are crushed or made into a puree. Crush soft fruits, which have been washed, sorted and pitted, with a slotted

Enjoying your bounty — freezer bags, preserving jars, drier

spoon. Or, for firmer fruit, use a food processor at the chop setting. Crushed fruit is good for puddings, ice cream and fillings. A blender set on puree, a colander or food processor will make a softer puree for jams, pies and cake fillings.

Prepared fruits are packed unsweetened, in sugar or syrup. To pack unsweetened fruit, simply pack it into containers whole or cut. Press it down into its own juices with waxed paper and close and seal the container. If the fruit does not have sufficient juice, you can add water containing dissolved ascorbic acid (vitamin C). This will also keep the fruit from discoloring.

If using the sugar pack method, cut up fruit into a shallow bowl and sprinkle sugar over it. Mix the fruit until a sugar syrup forms. As with unsweetened fruit, submerge the sweetened fruit pieces in their juice before sealing the container.

For the syrup pack, dissolve sugar in hot water. Three cups of sugar to 4 cups of water will make a medium syrup. Use more sugar for tart fruits and less for sweet ones. Cool the syrup before packing. Cover the fruit with the syrup and seal it into the container.

Your freezing container may be a jar with a screw-on lid, a freezer container, or a zip-lock freezer bag. The object is to lock out air and keep in moisture. Screw-on lids should have a rubber rim. Zip-lock bags may be sealed by heating the zip portion with an iron. Square plastic freezer containers are the best bet. They are easily stacked and space efficient.

A few tips. Never prepare more than a few quarts at a time; pack in family servings; and always leave expansion space, or headroom, in your container. Don't forget to label containers with the contents and date and always use them in chronological order.

Canning

Are you sure you want to do this? Canning can be a trial. When canning, you can either cold pack or hot pack (partially cooked). The cold-pack method is recommended for fruits because it preserves their shapes and fresh flavor best. Cold packing involves putting cold, uncooked fruit pieces into the canning jar and covering it with hot syrup or juice. A medium canning syrup may be made from 1 cup sugar to 2 cups water. Seal the jar and "process" it in boiling water to cover, or in a pressure cooker. Cooking fruit at below 2000 feet above sea level does not require a pressure cooker. Above 2000 feet above sea level, add 2 minutes to the cooking time per each 1000 feet.

The United States Department of Agriculture puts out an excellent home and garden bulletin on canning. You may order it through the U.S. Government Printing Office, if you are so inclined. Or, you might simply sit under your apple tree contemplating your harvest. Newton got some of his best ideas that way.

Papaya

Tree Sources

Where do new trees come from? Seeding, grafting, budding, layering and rooting. How do you begin? Grow trees yourself and learn how to graft, bud, layer and root. Or, purchase them in the state you desire from nurserymen and garden centers. Order them by mail from seed catalogs and fruit and nut associations or exchanges. Trade them among family and friends.

PIT POWER

Pits have potential. Who among you hasn't pondered the possibilities of pits? You've just munched down a peach or nectarine or halved an avocado. And there it is — the pit. Could this be the start of something big? Most definitely, *yes*. There are a few drawbacks to production, however.

Trees grown from seeds and pits may or may not produce fruit and nuts. If they do, the resulting fruit and nuts may be different from, if not inferior to, those produced by the parent tree. Trees will require longer to bear and may not be as resistant to diseases. Their size may vary from the parent tree also.

These factors were not a deterrent to the Rare Pit and Plant Council of New York City. The Council began 17 years ago, when a group of closet avocado-pit growers went public with a pit-growers contest. Since then, they've expanded, selecting many of their pits from fruits and nuts offered at ethnic markets. Debbie Peterson, cofounder, gave a number of tips for selecting productive pits.

Debbie says slightly overripe fruit is best. Once you've eaten the fruit, plant the pit immediately while it is still moist. As a starter packet, try dampened peat moss in a zip-lock bag. Leave the pit and packet on top of the refrigerator where it will receive bottom heat. Smaller seeds may be placed in small cubes of dampened peat moss contained in plastic netting to sprout. These are called *jiffy pellets.* When the tree has sprouted and grown to the height of 6 inches, remove the net and transplant the sprout into a larger container using potting soil. It's a special treat when trees grown from seed bear fruit. Celebrate the new arrivals.

PROPAGATION

Grafted trees grow faster and are more certain producers than trees that are not grafted. If you are a diehard do-it-your-selfer, you can do the grafting by yourself. Your local nurseryman will help you select the rootstock and scions best for your area. Except in the case of genetic dwarf trees, the rootstock will determine the size, resistances and yield, and the scion — a bud or shoot cut from another tree — will determine the size and quality of the fruit or nut. The scion is grafted or fused, bare wood to bare wood, to the rootstock. If you've grown your own 1- or 2-year-old seedling, you may use it as a rootstock.

You can whip graft, side graft, inlay graft, T-bud or chip bud. Depending on which method you choose, you'll need a grafting knife, exacto knife or razorblade, grafting tape and perhaps a hammer and nails. The general principles are the same for all grafting. Tapering cuts are made in the scion, be it a shoot with a bud or a bud on a chip of bark, to expose the cambium layer just below the bark. A "matching" cut is made into the rootstock. The two cuts are mated, locked together and held in place with nails or tape. When the budding section shows further growth in 5 to 6 weeks, the rootstock trunk is cut back with a heading-back diagonal cut to just above the union. The exceptions to this method are whip and inlay grafting where the scions are grafted onto rootstocks which have already been cut back.

Citrus, peaches, plums and pears do well with bud grafting. Nut trees flourish with inlay and whip grafting. In general, deciduous trees are best grafted during a dormant period and evergreen trees, such as citrus, in the spring. Warm temperatures and high humidity, such as you might find in a greenhouse, make for successful grafting.

There are several other ways to propagate from existing plant-ings. Take 8-inch terminal shoots from year-old wood of fig trees in late winter. Bundle them, turn them upside down in a trench, and cover them with soil until late April. The cut end of the shoot at the surface will thicken, forming the basis for rooting. Invert the shoots and plant them heads up, with 6 inches of shoot below the soil. They will take root and be ready for transplanting within a year.

You can root some cuttings directly from the tree. This works well for citrus. Choose a soft or semihard shoot about 8 inches long. Make a diagonal cut to remove the shoot and then cut off the lower leaves. Strip off the bark at its base and then dip it in a purchased rooting hormone powder. Root it in a container filled with peat moss and perlite. You may cover the shoot loosely with a plastic bag to maintain humidity. When the

 Bernie Bromka of Melbourne, Florida, deserves a medal for his propagating efforts. In 1980, while at the naval base in Rota, Spain, he made a military hop to Europe, where he discovered a tree bearing loads of fruit near the base exchange. He sampled some and tucked the pits away in his carry-on bag to avoid littering. Emptying his bag back home in Rochester, he found the pits and planted them. He transplanted his seedlings to his winter home in Florida. In about 4 years, they began to bear fruit. He planted their pits in a 4-foot by 4-foot tree nursery and the next year, gave away a dozen small trees to his neighbors.

Quite proudly, he says, "Now, in 1989, there are seven residents here at Lamplighter Village who have vigorous loquat trees as large as my original tree. One of my trees was planted in Fort Lauderdale. And two years ago, I gave three of them to the Brevard Rare Fruit Grower's Council for their plant auction. In a year or so, my tree will be a grandparent. It now gives me fruit twice a year, shade to sit under, and is a great topic of conversation for many of my snowbird friends."

shoot is well established and puts out new leaves, transplant it to a larger container, using potting soil.

In some cases, a shoot can be stimulated to produce roots while still on the parent tree. The method is called air layering. After selecting your shoot, peel away a ring of bark at the end to be rooted, and scrape away the light green cambium layer. Surround the cut with sphagnum moss held in place by plastic wrap tied off at the bottom to form a cup. Keep the moss moist to encourage the shoot to form roots. When a good root system has penetrated the moss, clip the shoot and plant your new tree.

PURCHASING AND EXCHANGING

Branched trees may be purchased at your local nursery. If there is no local nursery, search out gardening centers in retail stores. In many areas, chain department stores will have a special gardening area where tools, supplies and books can be purchased in addition to the trees themselves. Sears, Wards and K-mart come to mind. Building supply chains very often dedicate large areas to gardening pursuits and also offer materials to build patios, decks, greenhouses and planters, along with books, supplies and tools. Supermarkets and sometimes five and tens have seasonal offerings. Chains such as Franks and White Rose are dedicated to the home gardener's needs.

If none of these sources is readily available to you, you must be living in either a concrete jungle or the outer reaches of the U.S. Check with your library, library service or county extension for a listing of specialized nurseries. The appendices and bibliographies of gardening books also are good sources for locating mail-order nurseries. Or, write to the U.S. Government Printing Office for their directory of pamphlets on gardening.

The Mailorder Association of Nurseries in Laurel, Maryland, publishes a yearly edition of "The Complete Guide to Gardening by Mail," listing its membership and their specialties. The association is national, but a good percentage of members are regional nurserymen who are specialists on their local turf. This is a good resource.

Seed catalogs themselves are wonderful sources, giving a rundown on each of their offerings as well as planting instructions and other valuable information, such as how to design and build a windbreak. Some sell specialty supplies, tools and books. And they are up-to-the-minute on the latest developments. Have you heard of the filazel? Bear Creek Nursery lists several varieties of this relatively new cross between filberts and hazelnuts, which combines the hardiness of the hazel with the size and taste of the filbert.

The Miller Nursery in New York has been in the gardening family for 109 years. Their colorful planting guide and catalog offers dwarf and antique apples, dwarf pears and multiple-graft trees. Gardening gadgets and cookbooks round out their inventory.

Many nurseries specialize in one type of tree, such as the Lychee Tree Nursery, The Adams Citrus Nursery, The Fig Tree

Nursery, and the Chestnut Hill Nursery, Inc. See the appendix for their addresses.

ASSOCIATIONS AND CLUBS

Associations and clubs network among growers and spur the exchange of information. Many are also sources for books, tours, conferences, seed exchanges and listings of nurseries and clubs.

The Friends of the Trees Society works to preserve forests and plant trees. The Friends of the Trees' yearbook is a compendium of tree knowledge and lists nurseries, book publishers, clubs and conferences.

The excellent information and tips for the indoor grower once carried by the now-defunct Indoor Citrus And Rare Fruit Society newsletter appears now in *Fruit Gardener*, published quarterly by the California Rare Fruit Society, The Fullerton Arboretum, California State University.

If you're a nut for nuts, the Northern Nut Growers Association can supply you with informational newsletters and an annual report. They are gold mines of information on propagating and growing nut trees.

Many clubs and associations foster the exchange of seeds, scions and saplings. The Seed Savers' Exchange in Decorah, Iowa, known for its antique vegetable seeds, now has a fruit-tree scion exchange. Another group, The Seed People, holds an annual networking meeting in Montana for seed exchange purposes.

In fact, ever since humankind gave up the nomadic life, people have been getting together to trade seed and talk about their crops. Friends of the Trees sponsors such an event for modern growers. They call it "Seed and Plant Exchange Day." The public is invited to bring seeds and plants to exchange and/or sell, and the Friends also sell some of their own plants. The society is willing to show you how to organize a seed-and-plant day in your area. The organization also publishes the *Perennial Seed Exchange Want/Have Listing* which encourages tree propagation of all kinds.

Nuts

PECANS, WALNUTS, HAZELNUTS

It's all in a nutshell — monosaturated fat, protein, potassium and calcium. Nuts are a healthful addition to your daily diet. Now, the smaller trees and bush varieties greatly enhance edible landscaping.

Pecans

Characteristics. Pecan trees live to a ripe old age and grow to be very tall. We know of one tree which grew from a seed brought from Georgia to Florida in 1924. The tree is now 40 feet tall and still produces a yearly crop of tasty pecans in an area which has changed from a homesteaders' paradise to an industrial environment.

Planting Tips. Soak pecan seeds three days before planting them. Pecans are self-pollinating, but trees produce larger nuts when they are cross-pollinated. So, if you have space, plant two. There is no lack of variety. The USDA Field Station at Brownwood, Texas, recently released 14 pecan cultivars named after Indian tribes. If you are planting seedlings, bear in mind that grafted trees produce nuts in 6 to 8 years in the North and from 3 to 4 years in the South, while seedlings take about 10 years. When planting pecan trees in sandy soil, Master Gardener Jim Bidwell has found that adding four pounds of powdered zinc to the planting hole and mixing it with the soil hastens healthy growth and first nut production.

While Georgia, California, Florida and Texas are commercial pecan producers, home growers in mild climates can grow pecans. Some locations are better than others. Pecans may not ripen in Oregon, Washington, and Pennsylvania as well as they do in the mountains of Virginia and West Virginia.

The Missouri hardy pecan is suitable for Zones 5 to 9, the Surecrop pecan has survived 18 degrees Fahrenheit, and the

Hardy Giant has yielded a heavy crop after surviving a freeze of 20 degrees Fahrenheit. All are listed in *The Stark Brothers Catalog*.

Another choice is the Cape Fear pecan, for Zones 7 to 9. This long-lived tree bears nuts at a young age. Its kernels are well-formed and won't break when the shell is cracked.

The Major pecan, a hardy, grafted Midwestern paper-shell nut for Zones 6 to 8, bears almost perfectly round nuts with sweet plump kernels but needs cross-pollination. This tree can be planted with the Colby pecan for cross-pollination purposes in Zones 6 to 9.

Most folks now prefer thin-shelled pecans. You can crack their thin shells by pressing two of them together in your hand. The Mohawk variety does well in Zones 7 to 9. The Stuart pecan, also popular in the South, will grow well, producing nuts with easy-to-crack shells and delicious, easy-to-separate kernels.

One of the most lovely Southern belles, the Schley pecan, will supply cool, refreshing shade under its fast-growing limbs. When harvesting these plump, tasty self-pollinating kernels you will congratulate yourself for choosing a dual-purpose tree, one which provides you with food and natural air-conditioning. It produces even more nuts when pollinated with the Cape Fear variety in Zones 7 to 9.

If you have a small lot in Zones 7 to 9, plant the compact Cheyenne paper-shell pecan which bears heavy crops at an early age.

Our long-time favorite, but hard to find, is the large paper-shell pecan called the Kernoodle. It is harvested just in time for Christmas. This variety is listed as a Florida-grown pecan, but we order these special nuts from Georgia.

Care. Because of the long tap root, it is very important to keep your pecan tree well watered. Give it a bucketful of water every day for the first 2 weeks after planting. Continue watering every second day for 4 weeks. You can use stored rain water or recycled wash water very effectively this way. The Orange County Florida Extension Horticultural Division suggests planting pecan trees away from buildings, utility wires and other trees.

In addition to the space problems, pecan growers face the challenge of protecting their nut harvest from squirrels. If squirrels raid your tree, wrap a wide sheet of metal around the trunk. This metal shield will prevent squirrels from getting a toehold for climbing. Squirrels can be captured in a "friendly" trap and removed to a wild area. To help rid the garden area of squirrels, seal cracks and openings in houses and outbuildings to prevent squirrels from entering and nesting. Electrical fences are effective against marauding squirrels as well as people, but with burgeoning development and population, their use may be restricted.

Pecans need lots of nitrogen. They should be fertilized according to soil and leaf analysis. This service is available at most agriculture centers and should be used more by homeowners. Mature pecan trees need 30 to 50 pounds of 8-8-8 fertilizers per year, applied at intervals during the growing season.

Pecans require zinc for proper growth. Zinc can be applied as a powder or spray. Zinc sulfate (36 percent) makes a good foliage spray when mixed 1 tablespoon per 3 gallons of water. Make a first application after pollination, when the tips of the nutlets turn brown. Make two more applications at 3- to 4-week intervals.

If dusting, treat the soil by broadcasting powdered zinc from the tree trunk to the drip line. For trees up to 7 inches in diameter, apply 1 3/4 pounds for each inch of diameter. For trees over 7 inches in diameter, apply 3 1/2 pounds for each inch of diameter. Make sure to work zinc into the soil to prevent its washing away. Regular pecan fertilizers now come with 2 percent zinc added.

Caution: Do not use zinc sulfate 36 percent in spray or powder form if you are container gardening.

Harvesting. When pecans are ripe, they will begin to drop from the tree. Spread a cloth under the tree to catch them. If the tree is relatively small, you can hasten the harvest by shaking the tree. Children love to join in this fun. Larger trees are "picked" by beating the nuts free with a cane pole and dropping them on to the catchall, which is gathered into a sack. Nut flavor is best if pecans are hung in a mesh bag in a dry place for about 2 weeks after harvest. After you've finished your hard work, have a mulled cider and toasted nuts fireside.

Preparing. When you think of pecans, you probably think of pecan pie. Some of the old recipes require as much as two cups of sugar or syrup and as many as four eggs. Here's a recipe which will satisfy your sweet tooth and cut calories too.

Trees give us food, lumber, medicine and shade. They filter our water and process our air. Their beauty is food for the spirit. But as a valuable commercial commodity and as "obstructions" to development, they are an endangered species. Take heart, however. There does seem to be a growing international awareness of the importance of trees. Modern tree stories proliferate.

In the Reni Village in the Chamoli District of India, villagers embraced their trees to keep government workers from clear cutting them. That was in 1974. Today, the CHIPKO movement enlists villagers, students and social workers in a reforestation program. Over a million trees have been planted.

The government of Massachusetts has given away over 15,000 fruit trees to its citizens to plant in urban and country settings. The idea of planting food trees in public places was promoted by Clay Olson who inspired such programs as the Fruition Project which donates trees to schools, community centers and parks.

The Tree People of Beverly Hills planted a million trees for the Los Angeles Olympics in 1984. Their goal is to educate and motivate Californians to take better care of their environment. These people are activists and should be closely observed. They are out to turn our smog-filled cities green.

Low-Calorie Pecan Pie

1 cup chopped pecans	2 tablespoons melted butter
1/2 cup dark Karo syrup	A pinch of salt
1/2 cup sugar	1 teaspoon vanilla
2 eggs	

Mix all of the above together. Pour mixture into an unbaked pie crust. Bake the pie in a preheated 350-degree oven for 30 minutes. Cool it and enjoy!

Walnuts

Characteristics. Did you know that there are more than 15 varieties of walnuts growing in Asia, Europe and North America? In Brevard, North Carolina, during recent summers, I was amazed to see so many varieties displayed at farmers' markets. Walnuts grow in clusters of up to 10 nuts. They vary widely in color, shape and texture: the orange Persian walnut is light yellow; the Carpathian and English walnuts are tan; and the butternut is white. The heart-shaped walnut and kernel originated in Japan (*J. ailanthifolia*). In China, a huskless variety is grown; in France, the titmouse walnut sports a jacket so thin that small birds can pierce the shell and eat the kernel. Walnuts are pressed for their healthful oil in regions of Europe, and walnut hulls are used for their fish-stunning poison in Japan. The walnut tree produces a beautiful and durable wood used worldwide.

Planting. Walnut trees are found growing wild and cultivated all over the world, but they grow best in Zones 5-9. The trees present a special challenge to the grower with limited space. They can grow to 60 feet in height and two are needed for best pollination. Securing its place, the native American black walnut (*J. nigra*) exudes a substance from its roots which prevents other trees from growing within a 50-foot radius.

If you want to grow walnuts, reserve a special space for them. Many folks plant walnut trees on unused land, or in rows along the perimeter of their property or along their driveways. Walnut trees make excellent shade.

Bob Whitwood, who grows black and English walnuts on his farm in Illinois, recommends planting young trees in holes deep and wide enough to spread out their roots. Ample root space means better nut yield later on. He offers a special tip. When planting potted nut trees, wash off any sawdust or soil that clings to the roots. Debris clinging to roots may cause them to dry out and slow the growth of young trees. This tip applies in the planting of fruit trees as well.

Care. The Nogal, or subtropic black walnut, is so vigorous that it prospers when planted in highway dividers. The tree may be grown from seeds and will bear nuts in 8 to 12 years. Unplanted seeds lose viability quickly. If you are planting seedlings in a warmer region, chose *J. neotropica* rootstock for *J. Regia* walnuts.

Harvesting. Harvesting nuts requires some effort. Usually, nuts are picked up from the ground. Outer husks are removed and nuts are washed and dried and spread on drying trays. Commercially, this is done with warm air and fans. When properly dry, they are much easier to crack. Special heavy-duty nut crackers are available and worth buying. We've had a special Texas Nut Sheller for years and appreciate the double blades which make it possible to nip off both ends and remove whole nutmegs from the shell easily. Replacement parts are available from York Nut Sheller, Inc., San Angelo, Texas.

Preparing. Walnuts are a popular ingredient in cakes, cookies and candies. Here is a sweet that will become a favorite at your house.

Black Walnut Bars

2 eggs	1/2 teaspoon baking powder
1/2 cup sugar	1/2 teaspoon salt
1/2 cup flour	1 cup chopped black walnuts
1/2 teaspoon vanilla	2 cups finely cut dates

Preheat the oven to 325 degrees. Oil an 8-inch square pan. Beat eggs, adding sugar and vanilla. Combine flour, baking powder and salt. Stir this combination into the egg mixture. Add walnuts and dates. Spread the mixture evenly in the pan. Bake 25 to 30 minutes or until the top becomes dull. Cool. Cut it into bars. Remove the cake from the pan and dust lightly with powdered sugar.

Hazelnuts, grown mostly in the northwest U.S., were also part of early myth. Mercury's staff or winged wand is made of hazelnut wood and symbolizes communication and commerce. In early Rome, bridal torches of hazelnut wood lit the way to the marriage bed and blessed the happy union. Romans used divining rods of hazelnut to locate hidden water sources, minerals and treasure. In Norse myth, the hazelnut was the tree of wisdom, symbol of power and badge of honor.

Hazelnuts

Characteristics. The filbert or hazelnut originated in Turkey and is grown in Italy, Spain and the Willamett Valley of Oregon. It grows as a bushlike shrub but may be trained to grow as a small tree ideal for the patio. As a tree it can attain a height of 25 feet. It is an early and prolific producer, bearing nuts in under 5 years and continuing to bear crops in early fall for as long as 50 years.

Planting Tips. Filberts need cross-pollination and like a touch of cold. New varieties can tolerate temperatures as cold as -20 degrees Fahrenheit. The trees require slightly acid soil, well-drained and preferably fertile. Plant them a few yards apart where they will receive full sunlight. Cross-pollinate Barcelona with DuChilly or Royal for a truly tasty nut. Rootstocks resistant to filbert blight will produce smaller, less tasty nuts. Zones 5 through 8 are best for hazelnuts.

Care. Prune away suckers and shape your hazelnut or filbert in the open vase method for best results. This is the method commercial nut growers use. According to Dr. Dale Nash, of Brevard, North Carolina, a tree pruned in this shape rewarded him with 2 pecks of nuts in its second year. Hazelnuts are subject to worms, which bore holes in the nuts, and filbert blight, a black fungus which is usually fatal. Feed a sickly tree with nitrogen as recommended by your most convenient agricultural extension. Remove blight-infected branches with clippers sterilized in fungicide.

Harvesting. Hazelnuts will drop to the ground when ripe, but if you want to beat out animal marauders, pick them in early fall when they come away with a twist. Dry them in the sun before storing them in a cool place.

Preparing. Filberts are used in gourmet-type chocolate bars and work well in this recipe for glazed nuts. We used our homemade calamondin liqueur to make the glaze.

Liqueur-Glazed Nuts

1/2 cup blanched hazelnuts or filberts (walnuts and almonds may be used also.)

3 tablespoons liqueur

Pour liqueur into a 9-inch glass pie plate. Add nuts and stir to coat. Microwave on high for 4 minutes, taking out to stir at the end of each minute. They should be glazed and light brown when done.

Nuts in General

Almonds are grown commercially in California, but if you live in Zones 6 to 9, you could give this relative of the peach a try. There are 250 varieties. The tropical almond (*Terminalia catappa*) with its large leathery leaves has been planted widely in warmer parts of Florida and the Caribbean, but use caution in eating the nuts. Edwin A. Menningers' *Edible Nuts of The World* places a skull and crossbones beside this almond entry. Almonds are also very difficult to crack because of their hard layer of thick corky shell.

Chestnuts are making a comeback in the U.S. after being almost wiped out in a blight in the early 1900s. Chinese chestnuts are now being grown successfully in Zones 4 to 8, but grafted trees have not had a favorable survival rate in the warmer parts of the U.S.

Shelling these tasty chestnuts is not easy. The most effective method we have found to make them pop open is to cut an "X" on their flat surface, put them in a covered microwave dish with a little water, and cook them for 5 minutes. Cool them slightly. Avoid steam burns when you remove the cover. Remove the chestnut's shell and fuzzy covering while the nut is still warm. We have served chestnuts this way at parties. Each

 Ethel Fitzgerald, formerly of Harrisburg, Ohio, now of Florida, fondly remembers a hickory nut tree that grew in her backyard close to her kitchen window. She enjoyed it from first bloom to final bite in "wonderful white hickory nut cakes with nut frosting."

and eats his or her own. No fuss, no muss.

The macadamia nut, a member of the Protea family, is highly prized for its special flavor and is used in gourmet dishes. The nut grows predominately in Hawaii, South Africa, dry parts of Australia, and warm parts of California and Florida.

Pistachio nuts have been grown in Persia and Iran for hundreds of years. They were traditionally dyed bright red to hide the discoloration acquired during harvesting. Now that California has an industry producing natural-colored pistachio nuts, you can buy them almost year-round. While most of these nuts trees are grown in California and Arizona, producing trees have been reported in Merritt Island and Key West, Florida.

The pistachio nut tree is an attractive small tree which makes an excellent family fruit tree. Imagine the excitement of producing beautiful nut clusters with ivory shells in your own backyard. Pistachios are drought resistant and heat tolerant, but they need some winter chilling with temperatures below 45 degrees. There are several varieties with varied chill-hour requirements.

Pistachios provide 23 percent of the recommended daily allowance of thiamin, 17 percent phosphorus, 14 percent magnesium, and 12 percent protein. For more information on pistachio culture, contact the U.S. Cooperative Extension, 4145 Branch Center Road, Sacramento, CA 95827.

Plant now. The good that you do will live after you. Each year, Gladys Micheals brought homegrown Stuart pecans to our Rare Fruit Wine and Cheese party. Their flavor was superb, and her gift became a tradition. Now that Gladys has passed on, her husband and daughter continue to bring the gift of homegrown pecans to our gathering. It is somehow comforting and life affirming to know that the fruit and nuts we plant now will still be enjoyed after we are gone.

The next time someone says "nuts" to you, take it as a wish for your health and happiness, for that is just what these healthful, tasty kernels will promote.

CHAPTER 9

Citrus

by Gil Whitton

Everything we ever wanted to know about citrus was related to us by Gil Whitton, horticultural consultant, and well-liked host of radio and television gardening shows. Since we couldn't say it better ourselves, we decided to let him do it in this chapter especially dedicated to citrus.

Characteristics. Citrus is the queen of fruit. There are many varieties. The genus includes grapefruit (*Citrus paradisi*), small lime (*Citrus aurantifolia*), large lime (*Citrus latifolia*), lemon (*Citrus limon*), tangelo and murcott honey (*Citrus reticulata*), blood orange (*Citrus sinensis*), king (*Citrus nobilis*), satsuma (*Citrus unshiu*), citron (*Citrus medica*), kumquat (*Fortunella margarita*), and calamondin (*Citrus madurensis*). In addition to these, there are many species and crosses. For instance, the tangelo is a cross between a grapefruit and a tangerine. The limequat is a cross between a lime and a kumquat.

The Citrus Industry, an excellent reference book edited by Walter Reuther, devotes half of its 611 pages to naming the varieties of citrus. The sweet orange, for instance, is said to have originated in India and to have been brought to the New World by the Spanish. Its many varieties are in cultivation worldwide. We will be listing some of them along with other types of citrus under the "Harvesting" heading.

The citrus which we grow to eat is produced on an evergreen tree. It may shed a few leaves in the spring or when there is an insect or disease problem or after a freeze, but it is basically an evergreen. Some of the nonedible members of the species are deciduous, however. The *Poncirus trifoliata*, an understock on which citrus may be budded, is very cold hardy. The satsuma orange is generally budded on this understock. Satsuma will grow

Citrus — orange, grapefruit, lemon, lime, calamondin

as far north as Georgia and South Carolina in protected areas.

Understocks influence the quality of citrus fruit. They may make the budded part of the tree (a sweet orange or other type) sweeter, more disease resistant or more adaptable to various types of soils. The Citrus Experiment Station, at Lake Alfred, Florida, is constantly working with new understock in search of the perfect one.

There are varieties of citrus with few to no seeds, while others have many. The color of the fruit also varies. Identification of the fruit is determined by the seed count and the number of sections in the fruit, as well the color and texture of the skin. Of course, the type of leaf is also important in identification. Fortunately for us, most of the trees purchased are labeled correctly. It is fairly easy to tell the differences among a grapefruit, a sweet orange and a tangerine, or even a tangelo. Sour oranges (used as understock for budding) produce the most colorful fruit, a deep orange-red.

When John Horwath, of Fort Myers, Florida, lived in Santa Barbara, California, he grew a virtual Garden of Eden on his modest-sized property. "In the front yard I planted two almond trees, two macadamia, one Valencia orange, one feijoa (pineapple guava) trained as a small tree, and one elderberry bush. In the backyard I planted two Valencia oranges, one navel orange, one black fig, and one golden fig with extra large fruit — oh, so good! I also grew a loquat tree and several Concord grape vines as well as a vegetable garden. I harvested both nuts and fruit and even had a permit for winemaking."

Planting Tips. Citrus trees may be set any time, if they are in containers. It would be advisable to set them in late spring or during the summer, so the tree will have a full growing season to become established.

Most citrus is purchased in containers. The tree should be removed from the container before planting. Sound ridiculously obvious? Maybe, but some folks have set trees in the ground in pots! It is necessary to remove the pot.

Citrus trees should be set slightly high in the soil. In fact, the top of the soil in the pot should be about an inch higher than ground level. If the tree is a 3-gallon container size, add 3 to 5 tablespoons of fertilizer in the bottom of the hole and mix it with the soil. This will make the tree grow more rapidly. Water the tree as the soil is being placed around the ball. Eliminate air pockets which would prevent it from growing. New trees should be watered every day for a couple of weeks. After this, water 3 times a week for 2 more weeks. By this time, your new tree should have grown roots outside of the root ball and into the soil in which it was set. Water should be applied on a weekly basis thereafter for the rest of the growing season.

Spring is an excellent time to set trees. After the tree has been growing for 2 months, apply more fertilizer, using 6 to 10 tablespoons placed in the hole previously. Continue to fertilize every 6 to 8 weeks for maximum growth.

Care. Citrus trees should never be mulched. Grass should be kept away from the trunk at least a foot and even farther, if possible. This will help to prevent disease.

The young citrus tree may produce blooms and fruit the first year after planting; however, it is advisable to remove them and to give the tree another year to grow. If you cannot resist, leave one or two fruit to mature. The third or fourth year, you may want to leave a few more fruit on the tree, if the tree has at least doubled in size.

It should be mentioned here that citrus can be kept to a small size by proper pruning. If pruning is practiced, the dooryard grower will be able to have more varieties of trees in a limited space. Another method of having many varieties is to purchase rootstock with several different varieties budded onto it to make a single tree. Multigrafted trees may be difficult to find, and you may have to have a nurseryman graft one for you.

Insects may be a particular problem. New growth on the young citrus tree, as well as on older trees, may be subject to aphid damage. This very small green insect, if not controlled, may cause curling and twisting of the new growth. Safer (a form of concentrated soap), will give good control when applied according to the directions, as will tobacco juice. I soak cigar butts in water and spray with it.

White flies are small flying insects that lay eggs on the underside of the leaf. The adults do not feed on the leaf but the larvae do. A small yellowish-green shield structure develops after the eggs have hatched. It is the feeding of these larvae that injures the leaf. As the larvae feed, they excrete a sticky honey-dew substance. This falls on the leaves and black sooty mold, a fungus, feeds on it. Actually, the fungus does no harm to the tree except to limit sunlight absorbed by the leaf. An oil spray will control the whitefly larvae and cause the sooty mold to peel off the leaf. Once in a while you may notice some orange spots on the underside of the leaf. This is a friendly fungus (Aschersonia) which kills the white-fly larvae.

Black fly is also a problem in some areas. Although the two flies are almost identical except for their color, the black fly is somewhat more difficult to control. An oil spray should provide moderate control.

There are many types of scale insects which attack citrus leaves, limbs and even the fruit. All three of the pesticides discussed will provide control. An additional material, oil for citrus, is excellent, and it will also control spider mites which can be a problem. The modern-day oils used for citrus can be applied

Oranges have their allure. The great earth goddess Gaia was said to have given the orange or "golden apple" to Hera and Zeus on their wedding day. During the crusades, Europeans were captivated by the beauty of Saracen brides bedecked with orange blossoms which symbolized fecundity. They brought the custom home with them and made it part of their own wedding ceremonies. In the 11th century, under Moorish rule on the Iberian peninsula, oranges were forbidden to nonbelievers. Later, the orange was brought to the U.S. by the Spanish, grown in Catholic mission gardens in Florida and the Southwest, and propagated along the trails leading to them. Today, these are the two largest citrus areas in the U.S. Florida further encouraged citrus growing by giving homesteaders tax exemptions for planting citrus.

almost anytime, subject to a few limitations listed on the label. Always read the label on any pesticide before applying.

Chewing insects are at a minimum. The orange dog caterpillar, which is the larva of the giant swallowtail butterfly, is common on young trees. Use Bacillus thurinyiegsis.

Holes in grapefruit are caused by citrus rats. They feed on the seed of the fruit and will completely clean the inside of a grapefruit. Use a bait placed in the tree.

Rust mites feed on the outside skin of oranges, causing the fruit to turn a rusty brown color. The damage occurs mainly on the sunny side of the fruit. Rust mites can be controlled with an application of 2 tablespoons of oil emulsion (80 to 90 percent) in a gallon of water. This is a summer spray, mid-June to mid-July.

The length of time you must wait after spraying fruit before you can safely eat it has been established for a number of sprays. The label will specify these times on each bottle or package.

Prevalent diseases include Melanose (*Diaporthe citri*) a fungus which attacks the branches, leaves and fruit of grapefruit. The disease first appears as a dark speck on the leaf. As it progresses, it becomes rough and feels like sandpaper on the leaf, limb or fruit. It is mahogany in color and may appear as streaks on the fruit. The disease can be controlled by spraying the tree with neutral copper 3 weeks after the fruit is set.

Scab (*Elsinoe fawcetti*) attacks all parts of the tree. It is predominantly a problem on tangelos, but other citrus types can be affected. The disease produces a corklike growth on the limbs, leaves and fruit. It may begin to attack the leaves on the underside first, but it eventually works its way to the upper surface. The disease is distinguished from melanose by the corky growth. Neutral copper sprays provide control, if the tree is sprayed prior to new growth and then again after its petals fall.

Sphaeropsis knot (*Sphaeropsis tumefaciens*) causes large knots on the limbs. It attacks many types of citrus. The knots may occur at any age. They vary in size from very small up to 2 or 3 inches. Control is effected by pruning of the diseased limb well below the infection.

Stylar rot is a physiological problem that appears as a sunken area on the blossom end of the fruit. It is mainly a problem on limes. There is no known control.

Foot rot (*Phytophthora parasitica*) can kill a tree if not detected before it completely girdles the tree. The fungus is generally present in the soil, and when conditions are correct for infection at the base of the tree, it attacks. Mulching of trees, wet soil, grass around the trunk, nonresistant rootstock, or low budding of trees encourage infection. The first symptoms of the disease are small droplets of a gumlike substance on the bark at the base of the tree trunk, or near the bud union of the understock. The bark will also appear to be water-soaked. The disease will also attack the roots. Limbs may die on one side of the tree after infection occurs.

The following rootstocks are resistant to the disease: Swingle citrumelo, Trifoliate orange, Citrus macrophylla, and to a lesser degree, Carrizo citrange, sour orange, and Cleopatra Mandarin.

Control is difficult but the cultural practices mentioned help to prevent the disease.

Harvesting. The following listing includes most of the common varieties of citrus available from nurseries and garden centers, fruit description and maturity periods.

ORANGES, TANGERINES, TANGELOS, GRAPEFRUIT
Early Varieties Maturing September through November:

Parson Brown Orange	Fruit relatively large; 10-19 seeds, average number 14; peel rough or pebbly. Fairly coarse texture; deep yellow juice. Seeds run vertically; shape is short and dumpy.

Hamlin Orange	Fruit rather small, slightly oval; very smooth, thin, and fine-textured skin; 1-5 seeds. Used for juice.
Navel Orange	Mostly seedless; peel thick, smooth texture, medium to large fruit, to open rupture (navel) at blossom end. Used for eating.
Robinson Tangerine	Tangerine type, loose, easily peeled skin, smooth and glossy; medium to large size; large, hollow axis (center); segments 12-14; seeds 10-20 with green cotyledons. Tree thornless, dense, bushy leaf margin crenate (notched or scalloped).
Dancy Tangerine	Principal tangerine variety. Upright growth; small, narrow, pointed leaves. Fruit reddish-orange; loose skin, easily peeled.
Nova Orange	Longer, more narrow leaves. Fruit similar to Orlando variety with better color and flavor.
Satsuma Orange	Large, coarse, pebbly, loose-skinned fruit. Fruit matures before it turns yellow. Tree habit is viney, leaves small, veins stand out. This is one of the hardiest of edible citrus, especially when budded on *Poncirus trifoliata* rootstock.

Midseason Varieties Maturing December Through February:

Pineapple Orange	Usually round or slightly oblate fruit; 12-20 seeds; peel is deep orange color when ripe; rind firm, rather thick. Usually thornless. Seeds long, slender, run laterally. Juice or eating fruit.
Marsh Grapefruit	Large, white flesh, seedless, 2-5 seeds. Tree habit large and spreading, leaf, broad-tipped, rounded, with broad overlapping petiole. Fruit tends to be borne individually on small stems.
Ruby Red Grapefruit	Same as Marsh except red-fleshed with blushing through the skin.
Queen Orange	Similar to Pineapple except fruit has higher soluble solids. Juice and eating fruit.

Temple Orange	Narrow, pointed leaves; willowy, bushy-type growth; fruit slightly pear-shaped, many seeds (18-22); fruit sometimes shows navel; fruit on rough lemon understock is very rough and pebbly; on sour orange, it is quite smooth. Deep reddish-orange color. Most popular for eating. Freezes more readily than other citrus.
Duncan Grapefruit	White flesh, seedy fruit. Tends to cluster on rather heavy stems. Large spreading tree; leaves rounded with large overlapping petioles. May produce a light crop every other year.
Thompson Grapefruit	Same as Ruby, except that pink color does not blush through the skin or into the albedo.
Orlando Tangelo	Leaves have a peculiar curl and twist; very vigorous growth. Fruit is flat on bottom, very slightly necked with collar around stem. Few seeds. Peel smooth, thin. Early to mid-season.
Minneola Tangelo	Large, narrow, tapering leaves with dark green cast. Fruit very distinctly pear-shaped, with collar at stem end. Fairly smooth with thick rind, deep orange-red color. Superb flavor.
Murcott Honey Tangerine	Tangerine type, very upright tree growth. Fruit borne on end of limbs in exposed position. Skin easily peeled but firmly adheres to fruit. Seeds few to numerous. Color deep orange. Tends to bear a light crop every other year.

Late Varieties Maturing March through July:

Valencia Orange	Slightly oval, medium to large, deep orange when ripe. Few to not over six seeds. Juice.
King Orange	Mandarin type, deep orange peel and flesh color, 10-20 seeds, quite cold hardy; Large fruit with thick, rough peel. Peels easily. Excellent flavor.

LEMONS AND LIMES

Meyer Lemon
A favorite for dooryard plantings. It has a tendency to be a small bush with thorns, but will make a small tree. Bears fruit all year but mainly during winter. Fruit, medium to large, smooth skin, few seeds, light yellow in color.

Ponderosa Lemon
Fruit very large, low broad nipple, seedy, lemon yellow, slightly bumpy skin, medium thick, flesh pale yellowish green. Small tree good for patio planting. Very acid. Flowers purplish tinged. Fruits throughout the year. Thorny, leaves large. This tree is subject to freezing.

Rough Lemon
Used as an understock. Citrus budded on this understock and frozen may produce rough lemons as well as oranges. The understock grows more quickly than the budded orange. Fruit is medium size, irregular, with a furrowed collar or neck and broad apical nipple. The surface is deeply pitted, bumpy or rough. Thick skin is yellow to brownish-yellow. About ten segments. Flesh color is pale yellow to orange. Axis (center) is hollow. Many seeds.

Persian Lime
Also called Tahiti. Commercial lime of Florida. Fruit is medium to small with smooth, thin skin; green flesh color. Virtually seedless. Somewhat everbearing, mainly in winter. Tree vigorous, nearly thornless; purple color, usually faint in flowers and shoots. Leaves have small petiole.

Key Lime
Very small, rounded fruit, yellow when ripe, moderately seedy; greenish pulp color; highly acid with distinct odor of lime. Numerous small slender thorns. Small, blunt, pointed leaves; growth and flowers faintly purple tinged.

OTHER VARIETIES OF CITRUS

Calamondin	Matures October through January. Fruit is small, rounded, few seeds; thin rind; looks like small tangerine, upright growth, hardy, acid flavor.
Kumquat	Very small round or oval fruit, depending on the variety. Few seeds, fleshy, smooth surface, sweet-flavored and aromatic. Tree resembles calamondin in shape and size. The Meiwa variety is sweet. The Nagami variety is acid, used for preserves.

There are many other varieties of citrus. Those listed above are generally offered in the nurseries. A few retail outlets may have special varieties, such as the blood orange or sweet kumquat.

Citrus is great eaten fresh or juiced. Check the recipe section for other delicious ways to prepare it.

Unique Fruits

QUINCE, FEIJOA, CRABAPPLE

O' we've dined on quince and slices of feijoa
In the land were the palm trees grow,
But hot crabapple's a delight on a cold winter night
Where the trees are covered with snow.

Quince, feijoa and crabapple are three unique fruits which are tasty, nutritious and surprisingly versatile.

QUINCE

Characteristics. During October and November you may often find California-grown quince in your local produce departments. This cream-colored or light-yellow fruit comes in several shapes and resembles a misshapen apple or pear. It is best cooked.

Varieties include the apple quince tree, which is a heavy bearer. The fruit of both the apple-shaped and the Breczki quince varieties is excellent when cooked or stewed. Considered the best early-ripening fruit, the Portugal quince variety turns red when cooked. The Chinese variety has a number of virtues for the dooryard grower. The tree is a small deciduous tree which sometimes stays evergreen if planted in a protected area. It is lovely to look at in April and May when it bears solitary carmine blossoms on short spurs. The spurs are sprouted from year-old wood, and the fruits are egg-shaped and turn pale citrus-yellow when ripe. The Champion variety also produces greenish-yellow fruit.

Planting Tips. Quince doesn't mind a wee bit of frost on its cheeks and will do well in the Eastern United States in Zones 5 through 8. In cooler climates, a south wall will create a sheltered site for less cold-hardy types. Members of the Rare Fruit Growers, in Vero Beach, Florida, are experimenting with the quince in Zone

9. When choosing a planting site, keep in mind that a standard quince, although a slow grower, can attain a height of about 20 feet unless kept in a bushlike form by selective pruning. Generally, quince trees have an umbrella shape and low-level branches, all of which should be taken into consideration when choosing a site.

Care. Basically, the quince will grow in any reasonably fertile soil which is kept moist and protected against hard freezes. Often depicted in Oriental art, quince blossoms are both beautiful and fragrant. To encourage them and produce high-quality fruit, tip and thin out lead shoots of mature trees during the winter. Young trees need no special pruning. Quince trees may be propagated from cuttings and shoots, as well as by air-layering. Trees produced from seeds generally produce inferior fruit.

Harvesting. Use clippers and gloves to pick these easily bruised fruits before they are fully matured. Fully ripe quinces are yellow-orange. They will keep one or two months in cool storage.

Preparing. Unlike apples and pears, the raw flesh of the quince is hard and dry, and its musky odor may influence sweeter fruits. Do not allow quinces to mingle in the same refrigerator bin with apples and pears. When slowly cooked, quinces become soft and flavorful, and their color changes to pale apricot or carnelian.

Quinces are often cooked with orange juice or wine. They are sometimes combined with dried figs and, along with apples, can be made into cobblers and pies. One quince to 1 or 2 apples is about the right proportion to make a large merry-old-England-type tart.

Originating in Turkey and Persia, the quince made its way to Sussex where it grew wild in the English countryside and was sought after for use in wine- and marmalade-making. Quinces enhances the taste of roast chicken, turkey, pork or ham. Use your favorite apple pie recipe but substitute quinces for half the apples in the filling.

Baked Quince

4 large quinces	1/2 cup honey
1 cup coarsely chopped walnuts or pecans	1 teaspoon each cinnamon and nutmeg
1 cup raisins	1 teaspoon fresh ground ginger

Wash the quinces. Cut them in half and cut out cores. Fill the cavities with nuts and raisins. Add honey and sprinkle with spices. Bake at 350 degrees until fruit can be pierced with a sharp knife.

We grow our own ginger, and so can you. If you live in a warm climate, plant it outside. For cold climates, grow it indoors in a pot.

FEIJOA

Characteristics. A feijoa fruit resembles a puny avocado. Some folks think it tastes like a pineapple, and it is sometimes called a pineapple guava. But it is neither a guava nor a pineapple. It is a member of the myrtle family which is remotely related to the guavas.

The feijoa is a small, evergreen shrub native to South America. It has been grown successfully in Southern France, Southern California and New Zealand. Several attempts at growing feijoas in South Florida have been only partly successful. Commercially grown feijoas from New Zealand and Southern California are beginning to make their appearance at U.S. markets. Varieties include Sellowiana, Andre Besson, Choiceanna, Coolidge, Hehre and Superba. Also very successful are the triumph and mammoth varieties from nurseryman Harold Wright of kiwifruit fame.

Planting Tips. We've received reports that feijoas have survived in temperatures as low as 12 degrees Fahrenheit, but most become dormant at about 45 degrees. The feijoa may be planted in a broad range of zones and should be tried in more areas. It prefers dry rather than moist weather. If suckers and branches are limited, the feijoa forms a small tree of up to 20 feet in height perfect for container growing. It may be espaliered or grown as a hedge depending on space available. If your space is limited, choose self-fertile varieties such as the Coolidge, Edenvale or Mammoth which do not require cross-pollination.

Care. The feijoa can be started from seed, layering or cuttings, but seedlings will produce fewer flowers and fruit and mature later than those reproduced through cuttings and layerings. Some feijoas have borne fruit in as little as 2 years. The feijoa is salt-tolerant, requires at least 50 hours of chill and dislikes reflected sunlight and temperatures over 100 degrees. It prefers well-drained soil with an acid pH of 5.5 or higher. Fertilize it lightly, water it deeply, and mulch it. The feijoa is a relatively low-maintenance tree resistant to disease and pests.

Harvesting. Shake the tree and ripe fruit will fall at your feet, preferably onto a tarp which you can then gather up. The fruits may also be picked green and allowed to ripen indoors. Don't harvest more than you can use in a 2-week period. Beautiful, edible blossoms in spring are another advantage of this tree.

Preparing. Because the flavor varies with individual fruits, feijoas are probably best cooked or marinated in a mixture of water, salt, sugar and vinegar. They can be added to other tropical fruits in salads, sauces, souffles and ices. They are good company for apples and pears when baked in puddings and other desserts and make wonderful preserves.

Poached Fruit Compote

3/4 cup orange juice

2 tablespoons lime juice

1/3 cup water

1 tablespoon slivered orange peel

1/3 cup granulated sugar

2 oranges, peeled, halved and thinly sliced

3 feijoas, peeled, halved lengthwise, cut in 1/4 slices

1 cup melon or peach chunks

1/2 cup seedless red grapes, halved

Fresh mint garnish

In a medium saucepan, mix juices, water, orange peel and sugar. Bring to boil; boil rapidly for 2 minutes. Remove from heat. Stir in oranges, feijoas, melon or peach chunks, and grapes. Cover and refrigerate several hours or overnight. Serve garnished with fresh mint, if desired. Makes 4 servings. (Recipe courtesy Frieda's, Inc.)

CRABAPPLE

Characteristics. Related to the common apple, crabapples can be any small-fruited apple. The University of Wisconsin at

Madison has a large collection of crabapples. We have enjoyed seeing them in fruit during a fall meeting of the American Association of Garden Writers. The emphasis was on the use of crabapples as food for wildlife.

Blossoms are prolific and showy and the fruit, which ripens in the fall, may be about 2 1/2 inches in diameter. Bushlike varieties produce small apples while larger trees, ranging from 10 to 20 feet, produce larger apples. Varieties include the Dolgo crabapple which is considered the most hardy crabapple. Others to consider planting are Adams, Chestnut, Katherine, Montreal Beauty, Robinson, Whitney, Young American and more. The Centennial is the most cold hardy and will grow in the northern reaches of the U.S. The Whitney crabapple is light yellow with red blush or stripes and is the best for eating fresh and for pickling.

Planting Tips. Treat the crabapple as you would a standard apple tree. Mulch two feet from the trunk with straw or like material. See Chapter 3.

Care. Crabapples are susceptible to scab and fungal diseases, particularly the ornamental varieties. A dormant oil spray as buds open followed by a general-purpose spraying will ward off most pests. Check the "apple" entry in Figure 3 for details.

Do not allow a tree to bear fruit the first few years. Thin a heavy crop of apples before they mature to one or two per spur. Don't pull the fruit off. Use clippers or scissors. Prune the tree to balance growth and fruitfulness. See Chapter 4.

Harvesting. When fruit is ready to be harvested during the fall months, it will part from the spur with a gentle twist. For higher fruit, use a net on a cane pole.

Preparing. Bamboo steamers are wonderful for fruit and inexpensive. My son Bill bought one for us in Vancouver which we have used with great success.

During the summer of 1991, we found crabapples for sale at the local farmers' markets in North Carolina. Jack's brother, Earl Van Atta, shared some from his tree. We picked large bags of fully ripened apples the evening before the first frost of the year and made some beautiful low-sugar crabapple jelly. Note that the following recipe does not contain any pectin. Earl also made this crabapple jelly recipe using commercial pectin. Either way it's a treat.

Low-Sugar Crabapple Jelly

4 cups juice   1 cup sugar

Wash crabapples, cover with water and cook until soft. Pour apples into a cloth jelly bag and strain out juice. Measure into a large flat-bottomed kettle 4 cups juice and 1 cup sugar. Cook, stirring continuously, until mixture reaches the jelly stage (224 degrees on a jelly thermometer, or until the mixture "sheets" off the spoon). Our recipe yielded three 6-ounce glasses of beautiful red jelly.

To add spice to your plate try pickled crabapple.

Pickled Crabapple

3 pounds crabapple   cloves,
2 cups sugar   allspice berries,
1 teaspoon each: ground black pepper
 ground dry ginger,

Wash crabapples, remove their stems, and then steam the fruit until it is soft. Tie spices in a cheesecloth bag. Put the bag in a large kettle. Add vinegar, sugar and apples. Bring to boiling and let it simmer 20 minutes. Serve it with entrees to brighten color and taste.

CHAPTER 11

Super Trees vs. Monster Trees

One grower's super tree is another's monster tree. It all depends on which is the ideal tree for your site.

SUPER TREES

Since we're talking mainly about small sites — small plots, dooryards and containers — a super tree has to be small in size or capable of being easily pruned to a manageable size. Health concerns about pesticides, fungicides and fertilizers mean that super trees have to have top rootstock resistances to disease, pests and drought. They should be suited to the soil available without requiring too many amendments. If they are natives to your area, so much the better. These traits will make them less expensive to grow, because you won't have to buy large amounts of pesticides, fungicides and fertilizer.

Beauty is in the eye of the beholder. But who isn't attracted by an abundant display of fragrant, beautiful blossoms? Unless, of course, you're allergic to them or they attract swarms of unwanted insects. More blossoms mean more fruit — if the tree also happens to be self-pollinating. Otherwise, you must depend upon the vagaries of the winds and the insects to pollinate blossoms. Foliage has a beauty of its own. Are your trees deciduous or evergreen? An evergreen citrus in a container will brighten a gray winter room considerably. These are points to consider in selecting your super tree. But the most important is the yield and quality of fruit.

For containers, dwarf citrus is an excellent choice. They are evergreen and produce attractive fragrant blossoms year-round, if you keep them warm. And fruit ripens and will keep right on the tree. If you have only one indoor citrus, you may have to play the bee and hand pollinate by moving pollen from one blossom to another using a small paintbrush. This ensures that fruit will set. Candidates for indoor container super citrus are the dwarf Valencia and Temple oranges; the dwarf Star Ruby grape-fruit; the Eureka or Meyer lemon; the Mexican, Key or West Indian limes; and calamondin.

Dwarf fig trees are super deciduous choices for indoor containers. They are quite precocious and bear in their first year. If properly cared for, they will bear two crops. The Celeste variety tolerates relatively low temperatures and is sweet, juicy and dries well.

If you live in a cool climate and have a little more space, try "no ladder" trees — peaches, plums and pears — which produce more fruit earlier than the semidwarf. Or you might try a multiple-graft. Miller Nurseries offers a dwarf tree with five different grafts. There are several advantages here. The apples are self-pollinating. Varying maturity dates keep them in blossoms and fruit simultaneously for months. The trees come in "modern" or "antique" varieties. "Modern" trees produce Red

Meyer lemon in whiskey barrel

Delicious, McIntosh, Yellowgold Delicious, Northern Spy and Cortland apples. The "antique" varieties include Golden Russet, Snow, Summer Rambo, Roxbury Russet and Pound Sweet.

Pears are also available in multiple-graft family trees that are self-pollinating. And if you have an amiable nurseryman or are an adventurous type yourself, try grafting your favorite fruits to make a dwarf fruit-salad tree.

Nut trees are generally quite large and most are not recommended for containers or dooryards. The American filbert or hazelnut does come in a bushing variety that reaches a height of up to 10 feet. These are self-pollinating, have both flowers and catkins, and produce abundant nut crops in alternate years. They may be ground-layered, or rooted, and they are hardy to -25° F. Several planted together make an attractive hedge.

If you are in a dry hot area, a dwarf almond, which may reach a height of 15 feet, is a good choice, if you have the space.

MONSTER TREES

For our purposes, monster trees are large, spreading, sensitive to disease, difficult to pollinate, messy and require a large amount of time, materials and expense to maintain. Again, we are considering small plots, container and dooryard sites. In these settings, all standard trees may become monsters because they are too large for the site, difficult to harvest or may cause damage to surrounding structures. Most nut trees are extremely tall, some reaching 60 feet or more, and are difficult to pollinate. They do, however, make excellent shade trees, if you have the room.

A 15-year-old pecan tree can produce 100 pounds of nuts annually. Depending on what you intend to do with your harvest, that amount is either a blessing or a plague. English walnuts do not do well in the southern United States and are subject to blight, as are some chestnut strains.

In citrus trees, the grapefruit, as you might expect from its larger fruit, is the largest. Oranges and lemons follow in descending order. A mature grapefruit tree requires a space 25 by 20 feet. An ungrafted grapefruit may grow up to 40 feet tall. so, unless you're into grafting, stick with the store-bought varieties. Check your rootstock. The new Trifoliate Flying Dragon rootstocks cut tree size by as much as half but also change the tree shape somewhat, so that it is more bushlike.

The Friends of the Trees Society is sponsoring an organization known as Actinidia Enthusiasts whose goal it is to proliferate vine fruits such as the kiwifruit or Chinese gooseberry. Kiwifruits have caught on in the U.S. Their bright green fruit is a treat to the eye as well as to the tastebuds, and it goes well in dessert pastries. A kiwifruit growers association has been founded in California. There are more than 3000 acres of kiwi orchards in South Carolina, Georgia and North Florida.

Smaller is not always better. Small trees can be monsters too. Some dwarfs lose as much in quality as in size. Be careful in selecting dwarf peach and nectarines. Avoid trees grafted to either Western Sand or Nanking cherry rootstocks. These reportedly produce early but are short-lived, and their fruit is lacking in quality.

Disease resistance may also be cut by dwarfing. Pears grafted onto quince rootstocks are more susceptible to the dreaded fire blight. Their cold hardiness is also reduced. Trees that require two or three pollinators to bear fruit, such as many apples and pears, fall into the nuisance category in my book. Purchase self-pollinating trees whenever possible — unless, of course, you enjoy the role of matchmaker.

Recipes

APPETIZERS

Apple Compote

6 tart cooking apples   1 cinnamon stick (1-inch length)
1/2 cup sugar   1/2 teaspoon grated lemon rind
1 cup water
2 tablespoons white wine

Peel, core, and cut apples in half. Bring sugar, water and wine to boil; add cinnamon and rind. Stir in apples; simmer until just tender. With a slotted spoon, lift out apples. Boil syrup until thickened. Pour over apples. Sprinkle the top of each portion with an additional pinch of lemon rind. (Barbara Bell Matuszewski)

Toasted Butter Pecans

4 cups pecan halves (1 pound) 1/4 cup butter, melted
1 tablespoon seasoned salt

In a large bowl, combine pecans, salt and butter. Place the bowl in the microwave on high for 5 to 6 minutes. Stir it occasionally. Cool the nuts on waxed paper.

BEVERAGES

Mulled Cider

1 gallon apple cider   1 teaspoon grated orange rind
12 whole cloves   1/4 teaspoon ground allspice
5 cardamom seeds   orange slices studded with cloves
3 cinnamon sticks (1-inch length)

In a large, heavy pan, combine all ingredients except clove-studded orange slices. Cover and bring to a boil. Reduce heat: simmer 15 minutes. Strain. Garnish with orange slices. (Yield 1 gallon.)

Sparkling Apple Wine

16 pounds apples

2 gallons water

2 lemons

3 oranges

1/2 ounce yeast (2 packets)

Wash the apples. Do not peel or core. Cut away any damaged parts, and cut into pieces. Put into an unchipped enamel container.

Pour the water over the apples and bring to a boil, allowing the water to simmer until the apples are quite soft but not mushy.

Let the juice rest for one day, then siphon it off into a crock.

Wash the lemons and oranges. Cut off the rinds thinly and put them into the siphoned liquid. Squeeze the fruit and add the juice to the apple liquid.

Warm 1/2 gallon of the liquid and pour back into the crock so that all the liquid is lukewarm.

Dissolve the yeast in 1/2 cup lukewarm water and pour it into the liquid.

Cover well and let ferment for 1 or 2 weeks. When fermentation has stopped, skim off any scum from the top and strain the liquid into wide-mouth gallon jars. Let it stand for at least two weeks. You may cover the jars with cheesecloth. When the liquid is clear, it's time to bottle and cork.

Serve this wine very cold. When the bottle is opened, add a sugar lump and wait for it to dissolve before pouring. This added touch will make the wine sparkle for your guests.

Note: Fermentation may be speeded up by adding a teaspoonful of very dry sherry, brandy or vodka. Two eggshells per gallon will help the clarifying process.

SOUPS & SALADS

Cold Fruit Soup

1/2 lb. dried prunes

1/4 lb. dried apricots

1/4 lb. dried pears

2 qts. warm water

1/2 cup sugar

3 tart cooking apples, peeled, cored, diced

2 cinnamon sticks, (1-inch length)

2 tablespoons cornstarch

Wash dried fruit; soak overnight in warm water. Next day, add sugar, apples, and cinnamon. Cook until fruit is tender. Strain through a colander, reserving both fruit and liquid. Discard pits and cinnamon sticks. Pour liquid into a large saucepan. Mix cornstarch with a little cold water; add it to the pan. Bring the mixture to a boil, remove it from the heat, stir in the fruit, and let it stand until cool. Refrigerate until it is quite cold. Serve it in soup bowls, each portion garnished with whipped or light cream. Serves 8. (Barbara Bell Matuszewski)

(Note: Fruit soup is good as a dessert served over sweetened rice or vanilla ice cream. It also works well served over steaming oatmeal for a healthy breakfast pick-me-up.)

Nutty Chicken Salad

3 cups cubed cooked chicken
1 cup chopped pecans
4 hard-cooked eggs chopped
3 teaspoon chopped sweet pickle
1 1/2 cup chopped celery

2 teaspoon lemon juice
1/2 teaspoon salt
1/2 teaspoon pepper
3/4 cup mayonnaise

Combine first eight ingredients and add to mayonnaise. Chill and serve on lettuce leaves.

ENTREES

Feijoa-Chicken Curry

3 to 4 tablespoons olive/vegetable oil

4 skinned, boned chicken breasts, cut into bite-sized chunks or equivalent package stir-fry chicken

1 1/2 cups thinly bias-sliced carrots

1 cup cubed red, yellow, or green bell pepper

1 cup thinly sliced onion

1 clove garlic, minced

4 to 5 feijoas, peel, halve lengthwise and cut into 1/4" slices

1 to 2 tablespoons curry powder

1 teaspoon salt

1/4 teaspoon allspice

1/8 teaspoon pepper

1 teaspoon cornstarch

1 14 1/2-ounce can chicken broth

Hot cooked rice and condiments: chutney, raisins, shredded coconut, slivered almonds

In a Dutch oven, heat 2 tablespoons oil. Sauté chicken 3 minutes or until plump. Remove from skillet. Heat remaining oil in skillet; sauté carrots, pepper, onion and garlic for 5 minutes or until carrots are nearly tender. Add chicken back to skillet with feijoas, curry, salt, allspice and pepper. Stir cornstarch into broth; pour it into skillet. Bring mixture to boiling; reduce heat. Simmer it, covered, for 20 minutes. Serve it with rice and pass the condiments to sprinkle over each serving. Makes 4 servings. (Recipe courtesy Frieda's, Inc.)

Pork and Pear Chinese Dish

1 1/2 pound lean boneless pork

2 tablespoons vegetable oil

1 teaspoon ginger

1/4 teaspoon garlic salt

1/2 small onion, sliced

soy sauce

1 cup water

2 stalks celery, sliced

2 pears, peeled and sliced thin

1 (14-ounce) can bean sprouts

1 (8-ounce) can water chestnuts

1 (4-ounce) can mushrooms

2 tablespoons cornstarch

1/3 cup cold water

Brown thin slices of pork in oil. Sprinkle with ginger and garlic. Saute onions with pork. Add water, soy sauce, and cook, covered, until pork is very tender. Add celery, pears and additional soy sauce to taste, cover and continue simmering about 10 minutes. Add drained bean sprouts and water chestnuts to meat mixture. Heat. Meanwhile, stir cornstarch into cold water and pour mixture over all. Cook, stirring until sauce is transparent. Serve over rice. (Yield 6 servings)

VEGETABLES

Baked Squash and Apples

2 acorn squash
2 cooking apples
1/4 cup cashew nuts
1/4 cup maple syrup
1/4 cup butter, melted
2 Italian sausages (sweet or hot)

Cut squash in half crosswise. Take out seeds and stringy material. Dice peeled apples and slice sausages; fill squash centers. Over this, pour syrup, melted butter and nuts. Place squash in baking pan and cover with aluminum foil.

Bake at 350 degrees for 45 minutes. Spoon drippings over squash and stuffing. If extra apples and sausages are left, place in small baking dish and bake with squash. (Yield 4-8 servings.)

Vegetarian Tzimmes

1 pound medium carrots, diced
1 pound sweet potatoes, peeled and diced
1 cup pitted prunes (6 ounces)
1/2 teaspoon cinnamon
Grated nutmeg
3 tablespoons fresh lemon juice
1 tablespoon finely grated lemon zest
1/4 cup honey
salt to taste

Preheat oven to 325 degrees. Combine ingredients in a medium baking dish. Add 1/4 cup water and salt to taste. Bake covered until the vegetables are very soft, about 1 1/2 hours. Serve hot.

Whipped Chestnuts

1 pound chestnuts
1 tablespoon butter
pinch of salt
1/4 teaspoon pepper
2 tablespoons hot milk
1 cup finely diced celery

Shell and skin chestnuts. Drop them into salted, boiling water and cook until they are tender. Drain well. Mash them with butter, salt, and pepper. Add 2 or more tablespoons of hot milk. Whip until fluffy. Warm them over water in a double boiler. Before serving, stir in 1 cup of finely diced raw celery.

DESSERTS

Apple Crumble Tart

1 sheet frozen puff pastry
1 tablespoon unsalted butter or
margarine, melted
3 Golden Delicious apples,
peeled, cored and thinly
sliced
1/4 cup plus 2 tablespoons
all purpose flour.

1/4 cup plus 2 tablespoons
chopped, blanched almonds
1/2 cup sugar
1/4 teaspoon ground cinnamon
3 tablespoons unsalted butter
1/4 teaspoon vanilla extract

Thaw pastry 20 minutes. Unfold and cut into a 10-inch circle. Place on a large, ungreased baking sheet and brush with melted butter.

Arrange apples in center of pastry, leaving a 1/2-inch border of pastry. Mix flour, almonds, sugar and cinnamon in a small bowl. Add chilled butter and vanilla; cut together with a pastry blender until small crumbs form.

Spoon topping gently over apples, being careful it doesn't spill over the edges of the pastry and onto the baking sheets. Bake at 350 degrees for 30 minutes, or until golden brown.

Serve with Sabayon Sauce. Makes 1 large tart. (Yield 8 to 10 servings.)

Sabayon Sauce

3 egg yolks
3 tablespoons granulated sugar

1/3 cup wine (Sherry)

Place egg yolks and sugar in the top of a double boiler over simmering water. Beat mixture constantly with a portable electric mixer. When the mixture is foaming, add wine; continue to beat until mixture begins to thicken. Do not curdle. Makes 3 cups. Serve over apple crumble.

Tropical Apple Pie

Filling

2 cups sliced apples (substitute:
canned sliced apples)
2 tablespoons cornstarch
1 sliced banana
sprinkle of cinnamon

1 cup sliced papaya (substitutes:
peaches, pears, pineapple
or mangoes)
12 slices guava paste (Goya) or
homemade

Crust

Prepared 9-inch pie crust in Py-
rex baking dish

Crumb Topping

1 cup quick-cook oatmeal

1/2 cup wheat germ

1/2 cup oat bran

1/2 cup butter or margarine

1/2 cup dark brown sugar

Put the sliced apples and the cornstarch into a large bowl and mix. Put the mixture into the prepared pie crust. Place the sliced banana over the top and add a sprinkle of cinnamon. Add the papaya or substitute fruit and cover the mixture with sliced guava paste. Mix together the crumb topping ingredients and sprinkle over the pie.

Bake at 400 degrees until done — about 45 minutes.

Note: The original recipe for Tropical Apple Pie was created by Laurence Atkins and appeared in the Sarasota Rare Fruit & Nut Society Newsletter, Sarasota, Florida. It appears here with changes.

Maple Apple Crisp

3 pounds tart apples, peeled,
cored and sliced into 1/2
inch wedges

1 tablespoon fresh lemon juice

1/4 cup maple syrup

1/2 cup walnuts

1/2 cup all-purpose flour

1/2 cup whole wheat flour

3/4 cup light brown sugar

1 tablespoon cinnamon

1 1/2 sticks unsalted butter,
chilled and cut into 12 pieces

Preheat the oven to 375 degrees and butter an 8-inch square baking pan. In a large bowl, toss together the apples, lemon juice and maple syrup.

In a food processor or blender, combine the walnuts, all-purpose flour, whole wheat flour, brown sugar and cinnamon. Turn the machine on and off several times until the nuts are chopped and the ingredients are well mixed. Add the butter and continue turning the machine on and off until the mixture is crumbly, about 10 seconds.

Place 1 cup of the crumb mixture in the bottom of the baking pan and press it into an even layer. Bake for 10 minutes. Remove from the oven and arrange apple slices on top, adding liquid from the bowl. Sprinkle the remaining crumb mixture over the apples.

Bake the crisp 35 to 45 minutes, until the top is browned and the apples are bubbly and tender.

Pear Crunch Microwaved

2 teaspoons slivered almonds
3 medium-size ripe pears, chopped (peeled and cored)
2 teaspoons lemon juice
1/4 teaspoon almond extract

1 tablespoon all-purpose flour
1 tablespoon firmly packed dark brown sugar
1 tablespoon margarine
2 tablespoons regular oats, uncooked

Place almonds in a custard cup. Cover with heavy-duty plastic wrap, and microwave at high for 1 to 1 1/2 minutes or until almonds are toasted; set aside.

Combine pears, lemon juice, and almond extract in a medium bowl, tossing well. Divide among 4 (6-ounce) custard cups or individual baking dishes; set aside.

Combine flour and brown sugar; cut in margarine with a pastry blender until mixture resembles coarse meal. Stir in oats and reserved almonds. Divide among reserved custard cups.

Microwave at high for 6 minutes or until pears are tender, rotating cups after 3 minutes. (Yield 4 servings.)

No-Sugar Upside-Down Cake

Cake:

1 large egg
1/4 cup vegetable oil
2 tablespoons milk
1/2 cup unsweetened fruit juice (apple, orange, pineapple)

1 1/2 cups unbleached white flour
1/3 teaspoon baking soda
1 teaspoon baking powder

Topping:

2-3 cups chopped fruit
1 tablespoon vegetable oil

1/2 to 1 teaspoon cinnamon/
Dash nutmeg

Prepare topping by tossing together fruit, oil, and spice. Spoon mixture into an oiled 8-inch square casserole dish. Prepare cake by beating together egg, oil, milk, and fruit juice. Add flour, baking soda, and baking powder. Beat well. Pour batter over fruit. Smooth evenly into casserole dish. Bake at 350 degrees for 30 minutes. Cool until just warm and turn out onto a serving dish. Serve plain or with ice cream or whipped cream. (Yield 6 servings.)

Yogurt Pan Tart

1 cup flour
1/2 cup sugar
2 teaspoons baking powder
1/4 cup unsalted butter
1 cup nonfat yogurt

1 tablespoon unsalted butter
3 cups sliced fruit, including apples, plums, bananas, pears or nectarines

Topping:

2 tablespoons sugar
1/2 teaspoon cinnamon

Vanilla yogurt

Cream flour, 1/2 cup sugar, baking powder and 1/4 cup butter in a food processor until well combined. Add yogurt and process until smooth.

Melt remaining 1 tablespoon butter in a 9-inch square pan in the oven; swirl over bottom and sides of pan. Spread batter in the pan and layer fresh fruit in rows over the batter. Combine the remaining 2 tablespoons sugar and cinnamon; sprinkle over fruit. Bake in a 350-degree oven for 45 minutes until golden brown. Serve warm with vanilla yogurt spooned on top. (Yield 9 servings.)

BREADS AND PANCAKES

Apricot Loaf

2 large eggs
1/3 cup mashed banana
2/3 cup water
2 teaspoons vanilla extract
1/3 cup vegetable oil

2 cups unbleached white flour
1 teaspoon baking soda
2 teaspoons baking powder
2 cups finely chopped dried apricots

Beat together mashed banana and eggs until creamy. Add water, vanilla and oil and beat. Measure in flour, baking soda and baking powder. Beat well. Stir in chopped apricot until evenly blended. Spoon batter into an oiled and floured 9" by 5" loaf pan. Spread batter evenly in pan. Bake at 325 degrees for 45 minutes or until a knife inserted comes out clean. Cool completely on a wire rack before slicing. Serves up to 6.

June's Wholewheat Apple Pancakes

1 cup whole wheat flour
1/4 cup white flour
1/2 teaspoon salt
1 1/4 teaspoon cinnamon
2 teaspoons baking powder

1/2 cups milk, yogurt, buttermilk or sour cream*
2 tablespoons oil
1-2 eggs
1-2 apples, peeled and finely chopped

*(Buttermilk is by far the best. If buttermilk is used, use 1 teaspoon baking soda in place of 1 of the teaspoons of baking powder.)

Mix the first five ingredients in a large bowl. Add the remaining ingredients and stir lightly. Drop onto hot griddle. Serve with butter, cream, yogurt, jam, or syrup. (June Cussen)

PRESERVES

Before embarking on preserve making, test fruit juice for pectin content. Many fruits are found to contain enough pectin to make preserves without adding commercial pectins which require large amounts of sugar for jellymaking.

Pectin Test

1 tablespoon juice extracted from fruit

1 teaspoon sugar
1 1/2 tablespoons Epsom Salts

Combine ingredients and stir until salts are dissolved. Let stand for 20 minutes. If large flocculent particles (solid mass) are formed, juice has enough pectin to make jelly.

Quince Preserves

quinces
sliced, seeded lemons

sliced, seeded oranges
sugar

Pare quince, cut into eighths and reserve the peeling. Remove the cores and discard them. Weigh the quince slices. Cover the peelings well with water. Measure the water as you pour it over. To each quart of water allow 1 sliced seeded lemon and 1 sliced seeded orange. Cook these ingredients slowly until the fruit is tender. Strain the juice and add the quince slices to it. Cook them until they are almost tender. Add sugar. Allow as much sugar as the weight of the quince slices. Continue cooking the fruit. When it is tender, lift out the fruit and place it in jars. Boil the syrup until it is heavy. Pour it over the quince slices and seal.

Sour Orange Marmalade

sour oranges

sugar

Squeeze the oranges and cut the strips into chunks. Use a spoon to remove pith from rinds and discard pith. Slice rinds in thin strips. Place rinds in a pot and add water to cover. Boil for about 15 minutes. Add juice and boil for an additional 15 minutes. Cool it overnight. Then, place it in a larger pot. Add sugar in a volume equal to the fruit volume. Boil 20-30 minutes or until it sheets from the spoon. Put it in sterilized jars. Wipe jar rims, tighten bands, then invert for a moment to seal.

(Bill Bixby)

PARTY IDEAS

September through December when apples are freshly picked is an excellent time to have an apple-tasting party. This party idea was originated by the folks at Applesource, a supplier of fruit, plants and seeds located in Illinois. You can adapt it for your area. Here are the basics.

Sit-Down Apple-Tasting Party for Six

Choose 6 to 12 apple varieties. Several hours before guests arrive, remove the apples from the refrigerator so that they will be at room temperature for tasting. Wash and polish the apples and use computer or mailing labels to identify one of each variety. If you don't have labels, a folded index card may be used to identify each variety.

Plan to serve foods which will complement the apples and refresh the palate between samplings, such as natural cheddar cheese, fresh French bread and butter, or walnut halves. Also serve water. Alcohol dulls the ability to appreciate the nuances of aroma and flavor.

Use 4- by 6-inch index cards to prepare scorecards. The cards should contain places for the name of the variety, comments and score, as well as a scale for scoring:

Score Card

10	Absolute perfection
9	Best apple I ever tasted
8	Excellent
7	Very good
6	Better than average
5	Good
4	Fair
3	Edible but only if I'm hungry
2	One bite is too much
1	Aarrgh!

Set the table with a cutting board and knife for the host or "Master Carver." Set each place with a dinner plate, fruit knife and goblet. Be sure to include bowls for collecting cores and peelings. Position a pencil and a scorecard at each place.

When the guests arrive encourage them to admire the apple display before the tasting begins. Put out your complementary foods and fill the goblets with iced water. When the guest are seated, the host should identify each apple and tell a little about it. Using the cutting board, the host should then cut the apple into six slices and serve them to the guests. If you peel the apple before cutting it, peel it very thinly, since much of the aroma and taste are concentrated just below the skin.

Keep the pace slow and casual, allowing enough time for guests to evaluate and to record scores and comments. Repeat the process for each variety and have a scorekeeper tabulate points.

For larger groups, set out apples, complementary foods and water buffet style. Ask guests to bring small fruit knives and cutting boards and cut thin slices themselves. Tables should be set as before and individual score cards tabulated.

At informal tastings, there are no rigid rules. The point is to have fun and learn. Any apple that receives top scores deserves respect and attention. Unless you are planning a commercial orchard, the most important result is discovering which you like best. The appendix includes a chart to help you keep your apple-tasting record.

A large selection of apple varieties can be ordered from Tom and Jill Vorbeck at Applesource, RT.1, Chapin, Illinois 62628. Telephone 217-245-7589.

CHAPTER 13

GADGETS

Gadgets — wonderful appliances, devices and mechanisms, many of which you never knew existed but which you will find you can't do without, all contrived to help you achieve your gardening and cooking best — are gathered together in this chapter. In each instance, we paid particular attention to gadget effectiveness versus gadget cost. Go forth and equip yourselves.

GARDENING GADGETS

Soil Testing. The successful nurseryman must know his soil. You can avoid a trip to the county extension division if you analyze your soil yourself. Cherries and peaches need a neutral soil. Find out if yours is right with a pH meter. Simply insert the probe into the soil and the guage will read out the soil pH rating. pH meters cost from $15 to $20. Soil-testing kits which provide instructions for the pH, as well as the nitrogen, phosphorous and potash testing of your soil cost from $10 to $16, and the testing involves handling small amounts of chemicals.

Compost. Neatness counts in small patio gardens. A 12-cubic-foot vinyl bin with a hinged, tight-fitting cover and a trapdoor at the bottom for removing composted soil is quite compact, innocuous-looking and keeps the odor in for about $112. A two-handled, long-rod compost turner may be just the tool to aid in aeration at $15. More noticeable than the compost bin but more effective is the tumbler variety which consists of a perforated drum on an axle which can be rotated by crank, saving the back for more enjoyable exercises. You can buy a plan to build one of these for a total cost of under $30. Other aids to quick composting are compost tablets containing enzymes and micro-organisms which devour leaves, wood and garden debris. They come in packs of 4 for about $7.

Pest Handling. Insects — the good, the bad and the ugly — are a subject of deep concern for the fruit and nut grower. A $9 to $12 purchase of ladybugs is worth its weight in aphids, tree lice, scale, thrips, mealy bugs, boll worm and leaf worm. Ladybugs brighten up the garden, and if they alight on your sleeve, are said to bring good fortune and new clothes. At least that's what the old tale says. What's more, they vacate the premises when the job is done. Toads, too, put away a lot of insects. You can encourage toads in your garden by providing a toad house: an inverted clay bowl with a toad-sized door. A planter may be added to the top for a total cost of $20 to $30. And since toad habitat is under siege by acid rain, you will also be striking a blow for the environment by providing them with a home.

A bright red wooden apple-like sphere, hung in a tree and painted with Tangle stickem catches and kills insects dead for as little as $12.

If you are going the spraying route, a backpack-type sprayer, which holds about 4 gallons, weighs about 9 pounds and sprays out a quart of liquid in ten easy pumps, is on the market for $120. I prefer the trombone-type sprayer because it is more easily adjusted and cheaper.

Slugs will grow sluggish as they patronize the Slug Pub — a plastic dishlike affair with a cover which may be filled with beer. A Slug Pub without the beer will cost around $6. The device also provides a conversation piece and an opportunity for punning.

For nibbling rodents, use plastic tree wraps about 2 feet tall. They run about $3 for four. These will prevent nicks from mechanical devices as well. If these fiends are not put off by protective wrappings, try a humane trap. These are cage-type traps made of galvanized wire which, when baited appropriately, will attract and trap animals without harming them. A handle is useful in carrying the captives to remote areas or your most convenient animal control facility. Traps come in sizes from chipmunk to woodchuck and prices vary accordingly from $30 to $60.

For birds, try fine-meshed, fabric netting in sizes from 3 1/2 inches by 60 inches to 28 feet by 28 feet. Nets should be draped over the tree. Sunlight and rain penetrate the protective net but birds cannot. The price range varies from $8 to $30. A variety of commercial scarecrows are available ranging from an inflatable Wizard-of-Oz type, standing 6 feet tall, to a hard vinyl owl and inflatable snakes. These range from $5 to $30. Mr. Scarecrow may be more charming than effective in this instance.

The Cutting Edge. Sharp clippers, knives and shears will give you an edge in tending your mini-orchard. There are miniloppers and rachet hand pruners, as well as folding pruning saws and heavy-duty pruners. Rachet-built clippers require less pressure to operate, are lightweight and exist for the 3/4-inch twig or spur in intimate situations. These cost about $25.

Heavy duty pruners will cost around $30 and be heavier in weight and geared for in-close work but on slightly larger branches. They may have sap grooves and wire-cutting notches. Loppers are long handled and give the user the advantage of doing the job without getting entangled in the tree. These require a bit of strength to operate and are heavier in weight, but they can take on branches of about an inch in diameter. They run about $30. The folding pruning saw looks like a large pocket knife. It may be attached to a pole for those hard-to-reach limbs. The blade is designed to be clog-free so that it won't hang up halfway through the cut. The folding pruning saw will cost about $30. To keep all your cutlery honed, sharpeners are a must. These are specially designed for pruning implements and shears. Long-wearing silicon carbide Prune-mate and tungsten carbide Shear-mate retain the implement cutting angle accurately. Both come mounted for bolting to a workbench. They are available at about $13.

Harvesting. The nut gatherer is a fun machine resembling a slinking wheel on a pole, which, when rolled in an area where nuts have fallen, picks them up and keeps them within the hollow spiral. Cost is about $25. If you need to check the weather to see if the day will be conducive to harvesting, consult an Abnaki weatherstick from Maine. When hung outdoors, the weatherstick will predict the weather by moving upward for fair and downward for foul, a keen forecaster for only $8.

If it rains and the ground is a little soggy, just roll out your portable pathway made from all-weather cypress slats 1 inch by 14 inches and connected by rope binding for around $45. This walk is attractive and durable enough to make a permanent garden walk.

Gadgets — Apple parer, blender, pressure cooker, juicer, rachet nutcracker, baking dish

KITCHEN GADGETS

An apple baking dish pinions a cored apple on its central post and speeds its baking while catching the syrupy apple juice in its bowl. This cooking method creates a scrumptious dessert but at a price of $14 per baking dish.

For about $35 a cast-iron apple parer will clamp to your tabletop and peel your apple in spirals in only 5 seconds, or faster if you crank harder.

Ceramic birds act as cleverly concealed air vents for pies and may be poked through the crust to alleviate boil-overs and soggy crusts.

Pressure cookers, steamer/juicers and saucer/strainers are expensive but worth their weight in gold if you are living off the land. In pressure cookers look for cast aluminum, readable dials, extra safety latches, pressure safety valve and sturdy racks. The pressure cooker can be used for canning, meat dishes and sterilizing. Depending on the size and the number of canning jars they can contain, they range in cost from $95 to $120. A steamer/juicer reduces whole fruit to juice for $100. Six pecks of apples will make 9 quarts of juice and skins and seeds are left behind. The Back-to-Basics saucer/strainer costs only about $45 but you will have to use a bit of elbow grease in cranking up the press to produce a fruit puree. An oaken bucket press about a foot deep will yield 8 quarts of apple juice depending on the pressure you can exert. But it is authentic and charmingly rustic for $120. A centrifugal juicer makes great apple juice and may be used for carrots and pineapples as well.

Nuts! Who among you hasn't craved a better nutcracker after having crushed the kernel and littered the living room and your best sweater with nutshell fragments? For about $12 to $16 you can get a rachet-type nutcracker which is capable of leaving the kernel intact while exerting just enough pressure to crush the shell. These come in both the hand-held and vise-mounted varieties. For $45 you can get a Crax-All with which you can regulate the pressure and the result — whole or crushed kernels.

Take Note. The paperless notepad allows you to make grocery lists, recipes and schedules, attach your note to any metal surface and simply erase your message by pulling the erase bar across the writing surface, for about $8.

Peruse your seed and garden supply catalogs and magazine advertising for other handy instruments.

The Cook's Garden
P.O. Box 535
Londonderry, VT 05148

David Kay
One Jenni Lane
Peoria, IL 61614-3198
Telephone 1-800-535-9917

Gardeners' Marketplace
Storey Publishing
Schoolhouse Road
Pownal, VT 05261

Hastings
The Southern Gardener's Catalog
1036 White St., S.W.
P.O. Box 115535
Atlanta, GA 30310-8535

Miller Nurseries Catalog
Miller Nurseries
5060 West Lake Road
Canandaigua, NY 14424

Organic Gardening
Rodale Press, Inc.
33 E. Minor Street
Emmaus, PA 19098

Plow & Hearth
301 Madison Road
P.O. Box 830
Orange, VA 22960-0492
Telephone 1-800-627-1712

Stark Brothers' Nurseries & Orchards Co.
Louisiana, MO 63353.

CHAPTER 14

Tree Museums

Today, it seems that we are cutting down trees the world around. The disappearance of rainforests in Brazil, India and New Guinea is regularly featured in the news. Less exotic trees right here in the U.S. are being bulldozed for shopping malls, offices and condos on a daily basis. As more and more tillable land disappears under paving, we are also losing fruit, nut and other food-producing varieties as well.

These losses have not gone unnoticed. Over the years, conservatories, arboretums and gardens have been set up both privately and publicly to display, maintain and improve tree varieties. In a sense, these are tree museums. Many can be enjoyed free of charge.

Agricultural schools and universities or colleges with forestry and agricultural departments usually have display areas open to the public. State extension research farms are another possibility for the tree watcher. There are nurseries specializing in antique and rare varieties of trees. They hold open houses and arrange group tours.

Some tree museums specialize in one or two kinds of trees. Bear Creek Nursery in Northport, Washington, holds a spring open house to display its fine store of nut trees, including the hybrid filazel and trazel, as well as several kinds of antique apples. Or, you could visit the Miller Nursery in Canandaigua, New York, any time during business hours and view their fine collection of dwarf cold-hardy fruit and nut trees. They also have a collection of antique apples.

On your way to the Florida Keys, keep an eye out for the Redland Fruit and Spice Park in Homestead. Redland cultivates tropical and subtopical fruits, nuts and spices used for food and medicine. Also in Florida is the Marie Selby Botanical Garden in Sarasota. Their specialty is bromeliads but they also have many

Mango

other tropical and subtropical gardens including one of unusual edibles.

The Orcutt Ranch Park and Garden Center in Canoga Park, California, displays 13 acres of citrus as well as a 5-acre children's garden. Set the little guys loose and enjoy yourself.

There are tree museums with a religious scope. The Chinese Temple Garden of Oroville, California, contains trees with religious and mythical significance. The Three Sacred Fruits — citron, peach and pomegranate — are featured. The Sarasota Jungle Gardens of Sarasota, Florida, grows trees and plants mentioned in the Bible and also has an aviary of tropical birds. In New Jersey, the Beth Israel Memorial Park of Woodbridge contains trees and plantings commemorating those plants which sheltered, fed and/or provided medicine for the Israelites, planted in areas with names such as "Garden of the

Promised Land," "Garden of Moses," "Garden of Jerusalem," and "Garden of Kings."

Others promote trees native to their region. The Charles Huston Shattuck Arboretum at the University of Idaho Forestry School offers 200 species of trees native to the Northwest. Not to be outdone, the campus of the University of New Mexico sports native flowering shrubs, trees and plants, including cottonwood, locust and juniper. And Cathedral State Park in West Virginia has the largest store of virgin hemlock, black cherry and red oak in the East.

Historic gardens are abloom all over the country. Colonial Williamsburg has 100 theme gardens on 70 acres. All plants are either native or were imported over 200 years ago — replicas of the English Colonial period. A fruit garden is included. Thomas Jefferson's garden at Monticello in Virginia is as interesting as his architecture. Of special interest are the mulberry trees which were imported to encourage the growth of silkworms. National Colonial Farm of Accokeek, Maryland, has a collection of heirloom fruits set in an authentic colonial garden and offers a program in historical research along with pamphlets on colonial crops, husbandry and tools.

Independence National Historic Park, smack dab in the center of Philadelphia and adjoining the Independence Hall Mall, has several good examples of historic city gardens. Pemberton House Yard includes early fruit trees. Other nearby examples of early city gardens of the 18th century include orchards, vegetables and herbs.

The French Renaissance walled garden at the Biltmore House outside Asheville, North Carolina, displays an excellent example of espalier gardening which includes several fruits.

Did you know that John James Audubon, namesake of the Audubon Society, spent his first few years in the U.S. at the Mill Grove Estate near Valley Forge, Pennsylvania? While there he planted trees especially attractive to birds along 6 miles of trails. Today, they attract naturalists and gardeners as well as birds.

At Thomas A. Edison's winter home in Fort Myers, Florida, tropical plants, trees and shrubs from around the world are on display. Edison worked here on ingenious inventions such as making rubber tires from goldenrod.

Arbor Day, the day each year set aside by the U.S. government for tree planting, was begun by J. Sterling Morton, Secretary of Agriculture under President Grover Cleveland. That was in April 1872. You can tour Morton's estate, Arbor Lodge State Park and Arboretum, in Nebraska City, Nebraska. Morton planted many of the rare chestnuts, gingkos, tulip trees and pines himself.

If you're looking for a really extensive ramble through many kinds of flora, try the Strybing Arboretum in Golden Gate Park, San Francisco, which offers 1000 acres of gardens arranged in microclimates. A conservatory and nursery are located nearby.

Visits to San Simeon, William Randolph Hearst's former home on the California coast, require a reservation, but the harmony between the architecture and nature are worth the additional bother. The gardens display flowering fruit trees along a mile of trellised walkway. And an arboretum sports specimen trees such as acacias and eucalyptus.

Longwood Gardens at Kennett Square, Pennsylvania, is another extensive garden which offers 12,000 plants in landscape, arboretum and conservancy. It has been in existence since 1800 and sponsors educational activities as well as theatre events and fund raisers. Excellent seasonal exhibits are part of the show, as well as rare trees such as the Kentucky coffee tree. The gardens also have an orangery.

Founded in 1872, the Arnold Arboretum in Jamaica Plains, Massachusetts, has introduced 500 new plants. Today, it shows over 1000 trees and shrubs, including a 200-year-old conifer bonsai. Plant discoverer E. H. Wilson brought back many of the plantings along the famous Chinese Walk which you must have special permission to enjoy.

Fairchild Tropical Gardens in Miami requires a tram ride to properly view over 2,500 varieties of plants. Their rare plant conservancy and tropical trees are tops.

Hales Corners, Wisconsin, is the home of the 500-acre Alfred L. Boerner Botanical Gardens which supports more than 1,100 trees. The featured collection of dwarf fruit trees and a street-tree arboretum are worth seeing. Also in Wisconsin are the Mitchell Park Botanical Gardens of Milwaukee where three botanical domes house rare cocoa trees.

The University of Utah, State Arboretum, evaluates trees for forestry purposes and grows over 5,500 trees in more than 440 varieties. The trees are chosen for interest to botanists and gardeners.

The Native Plant Trail at Sandia Peak Botanical Gardens, Albuquerque, New Mexico, offers a view of native shrubs, trees, cacti and flowers. As hikers climb from 5000 to 7000 feet they encounter a variety of microclimates and plants.

And the list goes on. There are Shakespearean gardens, including vines, flowers and shrubs mentioned in Shakespeare's writings. Poets' gardens grow those plants — mostly flowers — often referenced in verse. There are medicinal gardens. And there are friendship gardens which contain plants donated by different cultures, countries and ethnic groups. There are gardens for the blind such as The Fragrance Garden for the Blind in Gary, South Dakota, which are planted for their fragrance, texture and amenability to touch. Many such fragrance gardens also offer recorded programs or descriptions for each type of plant displayed.

If this orgy of visual, olfactory and tactile treats is too much for you, you may want to approach your garden study in a more reserved manner. You might begin by contacting information sources such as The Horticultural Society of New York, the Pennsylvania Horticultural Society, the North American Fruit Explorers, or The International Dwarf Fruit Tree Association. I highly recommend to you *The Traveller's Guide to North American Gardens*, by Harry Briton Logan, a very helpful source for those seeking information on tree museums. *Gardens of North America and Hawaii: A Traveler's Guide*, by Irene and Walter Jacob, Timber Press, offers a brief but up-to-date guide to arboreta, botanic and historic gardens, conservatories, and even includes a list of open-house tours.

Now that you're inspired, get to work on growing your own family trees. We're all familiar with Adlai Stevenson's eulogy of Eleanor Roosevelt: "She would rather light one candle than curse the darkness . . . ." In the case of the container or dooryard growers, you will find it is better to plant one organic fruit or nut tree than to curse the pesticides at the supermarket. How about one tree a year?

Following is a list, by state, of the gardens mentioned in this chapter:

California
Cahoga Park: Orcutt Ranch Park and Garden Center
San Francisco Golden Gate Park: Strybing Arboretum
San Simeon: William Randolph Hearst Estate
Oroville: Chinese Temple Garden

Florida
Fort Myers: Thomas A. Edison's Home and Gardens
Homestead, Redland Fruit and Spice Park
Miami, Fairchild Tropical Gardens
Sarasota, The Sarasota Jungle Gardens
Sarasota, Marie Selby Botanical Garden

Idaho
University of Idaho Forestry School:
    Charles Huston Shattuck Arboretum

Maryland
Accokeek: National Colonial Farm

Massachusetts
Jamaica Plains: Arnold Arboretum

Nebraska
Nebraska City: Arbor Lodge State Park and Arboretum

New Jersey
Woodbridge: Beth Israel Memorial Park

New Mexico
Albuquerque: Sandia Peak Botanical Gardens
University of New Mexico Gardens

New York
Canandaigua: Miller Nursery

North Carolina
Asheville: Biltmore House

Pennsylvania
Kennett Square: Longwood Gardens
Philadelphia: Mill Grove Estate, plantings by
    John James Audubon
Philadelphia: Pemberton House, Independence Mall

South Dakota
Gary: The Fragrance Garden for the Blind

Utah
University of Utah, State Arboretum
Virginia
Williamsburg: Colonial Williamsburg
Monticello: Thomas Jefferson's Garden
Texas
San Antonio: San Antonio Botanical Garden
Washington
Northport: Bear Creek Nursery
West Virginia
Cathedral State Park
Wisconsin
Hales Corners: Alfred L. Boerner Botanical Gardens
Milwaukee: Mitchell Park Botanical Gardens

APPENDIX 1

Tree Sources

Directories of Nurseries

Brooklyn Botanic Garden
Nursery Source Guide
1000 Washington Avenue
Brooklyn, New York 11225

Mail Order Association of Nurseries
8683 Doves Fly Way
Laurel, Maryland 20707
(301)490-9143

National Gardening Association
Directory of Seed and Nursery Catalogs
180 Flynn Avenue
Burlington, Vermont 05401

Yankee Permaculture
TIPSY Directory
P.O. Box 202
Orange, Massachusetts 01364

Nurseries

Adams Citrus Nurseries, Inc.
2020 Dundee Road
Winter Haven, Florida 33880

Bear Creek Nursery
P.O. Box 411
Northport, Washington 99157

Chestnut Hill Nursery, Inc.
Route 1, Box 341
Alachua, Florida 32615

Bountiful Ridge Nurseries, Inc.
Box 250
Princess Anne, Maryland 21853

Columbia Basin Nursery
Box 458
Quincy, Washington 98848

Cloud Mountain Nursery
6906 Goodwin Road
Everson, Washington 98247

Edible Landscaping Nursery
Rt. 2, Box 343A
Afton, Virginia 22920

Henry Leuthart Nurseries, Inc.
Box 666
Montauk Highway
East Moriches, New York 11940

Hidden Springs Nursery
Rt. 14, Box 159
Cookville, Tennessee 38501

Just Fruit
Rt 2, Box 4816
Crawfordville, FL 32327

Miller Nurseries
5060 West Lake Road
Canandaigua, New York 14424

Monrovia Nursery Co.
18331 E. Foothill Blvd.
Azusa, California 91702

Patricks' Nursery
P.O. 130
TyTy, Georgia 31795

Stark Bros.
Nurseries & Orchard Company
Louisiana, Missouri 63353

Southmeadow Fruit Gardens
15310 Red Arrow Highway
Lakeside, Michigan 49116

Texas Pecan Nursery
P.O. Box 306
Chandler, Texas 75758

Wiley's Nut Nursery
116 Hickory Lane
Mansfield, OH 44905

NOTE: Local fruit and nut tree sources may be found in the market bulletin of your state agricultural department. Write to them for ordering information.

APPENDIX 2

ASSOCIATIONS AND CLUBS

American Pomological Society
Fruit Varieties Journal
103 Tyson Building
University Park, Pennsylvania
16892

American Society for Horticultural Science
701 N. St. Asaph St.
Alexandria, VA 22314-1998

California List of Rare Fruit
Clubs
California Rare Fruit Growers,
Inc.
California State University at
Fullerton
Fullerton Arboretum
Fullerton, California 92634

ECHO
(Educational Concerns for Hunger Organization)
174430 Durance Road
North Fort Myers, FL 33917

Friends of the Trees Society
P.O. Box 1466
Chelan, WA 98816

International Dwarf Fruit Tree
Association
A-338-A Horticulture Dept.
Michigan State University
E. Lansing, MI 48824

Living Tree Center
c/o Dr. Jesse Schwartz
P.O. Box 10084
Berkeley, California 94709

National Arbor Day Foundation
100 Arbor Avenue
Nebraska City, Nebraska 68410

National Gardening Associa-
tion
NGA Subscription Service
P.O. Box 52874
Boulder, Colorado 80322-2874

New Zealand Tree Crops Asso-
ciation
P.O. Box 1542
Hamilton, New Zealand

North American Fruit Explorers
Rt. 1, Box 94
Chapin, Illinois 62628

Northern Nut Growers
Association
9870 S. Palmer Road
New Carlisle, Ohio 45344

Rare Pits & Plant Council
C/O Debbie Peterson
251 West 11th Street
New York, New York 10014

Southern Fruit Growers
5715 W. Paul Bryant Drive
Crystal River, FL 32629

RARE FRUIT CLUBS

(Most rare fruit clubs give attention to all fruit crops growing in their areas, rare or not.)

Rare Fruit Council International
P.O. Box 561914
Miami, Florida 33256

RFCI Affiliates:
Brevard (RFCI)
P.O. Box 3773
Indialantic, Florida 32903

Indian River (RFCI)
P.O. Box 1117
Vero Beach, Florida 32960

Manatee (RFCI)
P.O. Box 1656
Bradenton, Florida 34206

Palm Beach (RFCI)
P.O. Box 16464
Palm Beach, Florida 33416

Sarasota Fruit & Nut Society
(RFCI)
4520 Camino Real
Sarasota, Florida 33581

Spring Hill Chapter (RFCI)
12333 Glen Haven St.
Spring Hill, Florida 34609

Tropical Fruit Club of Central
Florida
Lue Gardens
1730 N. Forest Avenue
Orlando, Fl 32803-1903

Tampa Bay (RFCI)
P.O. Box 20636
Tampa, Florida 33685

Non-Affliated Rare Fruit Clubs:
Florida:
Caloosa Rare Fruit Exchange
Box 3406
Palm Beach Boulevard
Ft. Myers, Florida 33905

Collier Fruit Grower's Council
P.O. Box 9401
Naples, Florida 33941

Rare Fruit & Vegetable Council
of Broward County
3245 S. W. 70th Ave.
Ft. Lauderdale, Florida 33312

Southwest Florida Rare Fruit
Growers Exchange, Inc.
P.O. Box 8923
Naples, Florida 33941

Tropical Fruit & Vegetable
Society of Redland
Fruit & Spice Park
24801 S. W. 187 Avenue
Homestead, Florida 33031

International:
Israel Rare Fruit
Ariel Shari
Horticultural Research &
Development
Jacobson 5 St. Rehovot
Israel 76206

Rare Fruit Council of Australia
P.O. Box 707
Cairns, Queensland, 4870
Australia

Appendix 3

PLANTING RECORD

Tree (Name/Rootstock/Source):

Date Planted:

Site/Condition:

Fertilzer

Watering

Pest/Disease
& Treatment

Yield/Date

Best Use:

Comments:

Notes: Site Plan

Appendix 4

Tips for the Organic Grower

Chart courtesy of Gardener's Supply Company, 128 Intervale Rd., Burlington, VT. 05401

Pests	Susceptible Trees And Damage	Where Pest Overwinters and Preventive Measures	Controls
Aphids (several species)	All fruit trees: Suck juices from leaves and fruit. Cause yellowing and tight curling of new leaves. Fruit discoloration can be caused by black sooty mold that grows on the honeydew secreted by aphids.	Eggs on twigs. Spray dormant oil during late dormancy as buds begin to swell but before any green shows. Apply only when temperatures are between 35 F and 85 F. Paint tree wounds to prevent feeding. Regularly loosen and cultivate soil 1-3" deep around trees to prevent root-feeding by woolly apple aphid. Aphids cause most damage in absence of beneficials.	Syrphid flies, green lacewings, ladybug larvae and predatory gall midges feed on aphids. Chalcid and braconid wasps parasitize. Tangle-foot spread in a band around each tree one foot above ground prevents ants from carrying aphids up trees. Safer's insecticidal soap or soapy water spray of 2 tablespoons dishwashing liquid/ one gallon of water repels some aphids. Spray entire plants. Rotenone dust as a last resort.
Apple Maggot	Apples, plums, pear. Females begin laying eggs in fruit 30 days after petals fall, larvae tunnel throughout fruit (versus codling moth larvae which tunnel from blossom end to center of fruit).	Pupae overwinter underground. Clean up dropped fruit. Remove nearby abandoned apple trees.	Monitor with red spheres smeared with Tangle-foot or use 4-6 spheres per tree to provide control in small orchard. Make bait traps to hang in trees: Solution of 1 part molasses diluted with 9 parts water and little yeast; or 2 teaspoons of household ammonia and small amount of soap powder in 1 quart of water. Hang plastic cups of bait on sunny side of tree about 60" high. Renew bait after rain.
Borers (several species)	All fruit trees, especially peaches. Borers drill holes in trunk or branch tips, leaving piles of sawdust or excrement near hole. Trees weaken and die.	Larvae under bark. Remove broken, diseased, or dead wood and any infested branches. Treat all wounds promptly. Coat trunk and large lower limbs with SilkaBen or whitewash or slurry of rotenone and diatomaceous earth to prevent egglaying in early spring, midsummer and fall. Monitor peach tree borer with pheromone traps.	Kill borer by very carefully probing into holes with wire and paint over wound with Lac Ealsam or other wound healer. Paint bark with interior white latex paint. Wrap bottom 12" by 18" of trunk with glossy magazine paper in April/May and remove and destroy in fall.
Codling Moth	Apples, pears, apricots, plums. Females begin laying eggs in leaves near fruit at petal fall. Grubs tunnel through apples from blossom end to center of fruit (versus apple maggot which tunnels throughout fruit), leaving excrement outside holes.	Larvae in cocoons under bark scales, debris or ground litter or a base of smooth-barked tree. Tie strips of burlap or corrugated cardboard around base of trunk and large limbs to trap larvae moving down tree to pupate. Then burn strips. Or cover strips with Tanglefoot. Tie paper bags around individual fruits after thinning in May or June. Carefully scrape loose bark from trees and remove ground litter. Larvae like to pupate in clover stems so remove any nearby. Monitor with pheromone traps.	Mating disruption. Trichogramma wasps parasitize moth eggs. Ryania, Pyrethrum, Imidan.

Disease	Susceptible Trees and Damage	Where Disease Overwinters And Preventive Measures
Apple Scab (Fungus)	Apples. Scab-like lesions on fruit and foliage. Deformed fruits, warped leaves, weakened trees. Worse during damp or rainy weather.	In fallen leaves. Resistant varieties. Clean up leaves under trees. Prune trees to open canopy and improve air circulation. Select site with good air drainage. Foliar feed with urea in fall to hasten leaf drop and decomposition and thus number of spores. Dormant oil. Sulfur or other fungicides, lime-sulfur, Bordeaux mixture sprayed prior to rain.
Black Knot (Fungus)	Peaches, plums, apricots, nectarines. Fruit rot, leaf spot, and cankers.	In cankers, old prunings and fruit, and on wild hosts. Remove and burn dead and infect branches. Dormant pruning can reduce susceptibility. Fungicides.
Cedar Apple and Quince Rusts (Fungi)	Apples, quinces, cedars. Orange or rust-colored leaf and fruit spots. Can cause defoliation and ruin fruit.	Cedar apple overwinters in galls on eastern red cedar. Quince overwinters on creeping junipers. Resistant varieties. Remove nearby junipers and red cedars. fungicides.
Cherry Leaf Spot (Fungus)	Cherries. Leaf spots that can eventually defoliate tree. Worse in damp weather. Spread by wind.	In fallen leaves. Remove fallen leaves. Bordeaux mixture, sulfur, lime-sulfur, other fungicides.
Fireblight (Bacterial)	Apples, pears. Branches wither and turn black. Orange-brown pustules on trunk are signal. Worse in high humidity.	Cankers under bark. Resistant varieties. Remove suckers and infected branches during dormant season. Sterilize pruners. Seal new wounds. Lush growth is more susceptible so don't overfertilize. Remove alternate hosts, i.e. wild apples, hawthorns, mountain ash, cotoneaster, control pear psylla, leafhoppers, and aphids which spread disease. Antibiotics. Copper sprays.
Peach Leaf Curl (Fungus)	Peaches. Curled, thickened leaves.	On twigs. Cut out and burn infected parts. Sterilize pruners. Fungicides.
Powdery Mildew (Fungus)	Many fruits. Leaves distorted and covered with white powder. Infected blossoms don't set fruit. Causes russetted fruit.	In leaf and flower buds. Resistant varieties. Prune to open canopy and improve air circulation. Cut out and burn infected tips. Sterilize pruners. sulfur and other fungicides.
Sooty Blotch and Fly Speck (Fungi)	Apples. Sooty blotch causes brownish green blotches on fruit that can be removed by rubbing. fly speck causes black shiny dots on fruit. These diseases often occur together. Worse in high humidity.	On twigs. Prune to open canopy and improve air circulation. Sterilize pruners. fungicides.
Wilts (Bacteria)	Many Fruits. Leaves wilt and die. Can cause branch dieback. Symptoms can resemble those caused by drought and girdling, so confirm diagnosis.	In soil. Keep fruit trees far away from vegetable garden to reduce spread. Avoid planting new trees in soil that might harbor wilt bacteria. Avoid overfertilizing with nitrogen. Cut out and burn disease wood. Sterilize pruners.

Pest	Susceptible Trees And Damage	Where Pest Overwinters and Preventive Measures	Controls
Green Fruitworms (several species)	Many fruits. Larvae eat leaves, make large holes in fruit causing drop. Fruits that mature are misshapen with corky areas.	Pupae or adults underground; eggs on twigs or leaves. cultivate under trees 1-4" deep. Monitor with pheromone traps.	Trichogramma wasps. Bt, Imidan
European Apple Sawfly	Apples. Females lay eggs in fruit from time buds turn pink to after bloom. Larvae burrow in fruit causing long curing russeted path on skin.	Larvae on ground. Remove ground litter.	Visual traps (white sticky rectangles).
Oriental Fruit Moth	Favor peaches and other fruits. Causes most damage in May. Larvae feeds inside twigs causing wilting and fruit causing black blotches.	Larvae in tree bark or ground litter. Plant early maturing varieties. Cultivate soil around trees 1-4" deep 1-3 weeks before bloom. Monitor with pheromone traps.	Mating disruption. Dust impregnated with mineral oil (viscosity of 100): sulfur 60%, 300 mesh talc 35 %, light grade mineral oil 5%; all percentages by weight. Apply at 5 day intervals 20 days before picking fruit.
Leaf Rollers (Several Species)	Many fruits. Larvae feed on buds, fruit, leaving holes and burrows in fruit and rolled leaves.	Larvae or eggs under bark on twigs, or pupae in debris. Remove debris and loose bark. Monitor with pheromone traps.	Pick off old leaves and destroy larvae. tichogramma and other parasitic wasps. Bt, dormant oil, ryania.
Plum curculio	Many fruits. Adults feed on fruit making crescent shaped cuts; larvae tunnel in fruit and cause drop.	Adults overwinter in soil and other debris. Collect and destroy dropped fruit frequently to eliminate larvae. Chickens roaming beneath trees eat adults cultivate soil around trees in late spring/early summer to destroy larvae and pupae.	Lay tarp beneath tree and bang tree to dislodge and collect adults. Rotenone/pyrethrum, Imidan.
Rodents and Rabbits	All trees. Chew on bark; can girdle and kill tree.	Keep ground cover closely mowed beneath trees. Move mulch away from trunks in gal. Trunk guard made by wrapping tree with old window screening. Push into ground few inches. Will work when no snow on ground. When snow up to 18", use screening and also wrap with tar paper or newspaper above it. Place 6" of gravel or crushed stone around trunk, extending 12" to 14" out from trunk. For rabbits: Paint trunks and major branches up to 30" high trees with undiluted liquid lime sulfur. Pain late fall, early winter.	Vitamin D3 bait kills them.
Tarnished Plant Bug	Many fruits. Feed on developing budgs shortly after new growth in spring. Damage terminal shoots and cause sunken areas on fruit.	Adults under leaf litter, stones, bark. Remove weeds beneath trees. Destroy infested fruit. Prune damaged twigs.	Visual traps (sticky white rectangles). Sabadilla.

Appendix 5

APPLE-TASTING RECORD

NAME	DATE/PLACE	APPEARANCE	TEXTURE/TASTE	USES	COMMENTS
Arkansas Black					
Cortland					
Delicious (Red/Golden)					
Freedom					
Golden Russet					
Granny Smith					
Grimes Golden					
Johnathan					
Liberty					
Lodi					
McIntosh					
Northern Spy					
Pink Pearl					
Pound Sweet					
Roxbury Russet					
Snow					
Winesap					

X	NAME	DATE/PLACE	APPEARANCE	TEXTURE/TASTE	USES	COMMENTS

THE GOLDEN HARVESTER
PLANTING CERTIFICATE

May the beauty of your blossoms
be exceeded only by the bounty of your table.

BE IT CERTIFIED THAT

In honor of_____, marking the occasion

of_____, this_____, this_____ tree is planted

at_____._____ day

_____ *(Signed)*

_____ *(Witnessed)*

_____ *(Witnessed)*

TREE - BUYERS GUIDE

TREE	HARD. ZONE	HOURS CHILL REQUIRED	HEIGHT	POLLINATORS	DISEASE, PESTS, SUSCEPTIBILITIES
ALMOND	6-9	400 HOURS	STAND. 30 FT. SEMI. 24 FT. DWARF 14 FT. MIN. 8 FT.	MOST REQUIRE CROSS-POLLINATION	OAK ROOT FUNGUS, CROWN ROT. CROWN GALL, BACTERIAL CANKER, ROOT-KNOT NEMATODES, DROUGHT, SHOT HOLE, RHIZOPUS FUNGUS, NAVEL ORANGE WORM, BIRDS
APPLE	3-9	400 HOURS	STAND. 30 FT. SEMI. 18 FT. DWARF 12 FT. MIN. 6 FT.	MOST REQUIRE CROSS-POLLINATION	CROWN ROT, WOOLY APPLE APHID DROUGHT, CODDLING MOTH, MITES, SCALE
APRICOT	5-9	400 HOURS	STAND. 24 FT. SEMI. 15 FT. DWARF 8 FT. MIN. 4 FT.	SELF-POLLINATION	PEACH TREE BORER, OAK ROOT FUNGUS, BROWN ROT, CROWN GALL, BACTERIAL CANKER, ROOT-KNOT NEMATODES
ASIAN PEAR	5-9	400 HOURS	STAND. 40 FT. SEMI. 20 FT. DWARF 15 FT.	REQUIRES CROSS POLLINATION	FIRE- BLIGHT SUSCEPTIBLE
AVOCADO	10	0 HOURS	STAND. 40 FT. DWARF 10 FT.	SELF-POLLINATION	CROWN ROT, DROUGHT
CITRUS	9-10	0 HOURS	STAND. 25 FT. SEMI. 20 FT. MIN. 6 FT.	SELF-POLLINATION	COLD HARDINESS, CROWN ROT, ROOT-KNOT NEMATODES, DROUGHT CITRUS BLIGHT, TRESTEZA VIRUS
EUROPEAN PEAR	4-9	400 HOURS	STAND. 40 FT. SEMI. 20 FT. DWARF 15 FT.	REQUIRES CROSS-POLLINATION	OAK ROOT FUNGUS, CROWN ROT. ROOT-KNOT NEMATODES, PEAR-ROOT APHIDS. GOPHERS AND MICE, FIRE BLIGHT
EUROPEAN PLUM	4-10	0 HOURS	STAND. 20 FT. SEMI. 14 FT. DWARF 12 FT. MIN. 8 FT.	SELF-POLLINATION	PEACH TREE BORER, GOPHERS AND MICE, OAK ROOT FUNGUS, BROWN ROT. CROWN GALL, BACTERIAL CANKER, ROOT-KNOT NEMATODES

(TREE BUYERS GUIDE - CONTINUED)

TREE	HARD. ZONE	HOURS CHILL REQUIRED	HEIGHT	POLLINATORS	DISEASES, PESTS, SUSCEPTIBILITIES
FIG	8-10	0 HOURS	SHRUB 10 FT. TREE 30 FT.	SELF-POLLINATION	HIGH RESISTANCES
FILBERT (HAZELNUT)	4-8	600 HOURS	SHRUB 4 FT. TREE 25 FT.	REQUIRES CROSS-POLLINATION	FILBERT WORM, EASTERN FILBERT BLIGHT, WESTERN BACTERIAL BLIGHT
NECTARINE	5-9	400 HOURS	STAND. 20 FT. SEMI. 14 FT. MIN. 4 FT.	SELF-POLLINATION	PEACH TREE BORER, OAK ROOT FUNGUS, BROWN ROT, CROWN GALL, BACTERIAL CANKER, ROOT-KNOT NEMATODES, MITES GOPHERS AND MICE
PEACH	5-9	400 HOURS	STAND. 25 FT. SEMI. 14 FT. MIN. 4 FT.	SELF-POLLINATION	PEACH TREE BORER, OAK ROOT FUNGUS BROWN ROT, CROWN GALL, BACTERIAL CANKER, ROOT-KNOT NEMATODES, MITES, GOPHERS, MICE, PLUM CURCULIO
PEAR	4-9	400 HOURS	STAND. 40 FT. SEMI. 20 FT. DWARF 15 FT.	REQUIRES CROSS-POLLINATION	FIRE BLIGHT, CODDLING MOTHS, MITES, PEAR PSYLLA, PEAR SLUG
PECAN	6-9	400 HOURS	STAND. 100 FT.	MOST REQUIRE CROSS-POLLINATION	PECAN SCAB DISEASE, APHIDS, PECAN NUT CASE BORER, PECAN WEEVIL
SOUR CHERRY	4-9	400 HOURS	STAND. 20 FT. SEMI. 15 FT. DWARF 10 FT. MIN. 8 FT.	SELF-POLLINATION	BIRDS, FRUIT FLIES, PEAR SLUGS, BACTERIAL LEAF SPOT
SWEET CHERRY	4-9	400 HOURS	STAND. 35 FT. SEMI. 25 FT. DWARF 15 FT. MIN. 8 FT.	MOST REQUIRE CROSS-POLLINATION	PEACH TREE BORER, OAK ROOT FUNGUS, CROWN ROT, CROWN GALL, BACTERIAL CANKER, ROOT-KNOT NEMATODES, VERTICULUM WILT, BIRDS
WALNUT	3-9	400 HOURS	STAND. 80 FT.	SELF-POLLINATION	ARMILLARIA ROOT, CROWN ROT, CROWN GALL, ROOT-KNOT NEMATODES, SALINE CONDITIONS, BLACK LIME DISEASE

SELECTED BIBLIOGRAPHY

BOOKS

Bailey, H. L. and Staff. *Hortus Third*. Macmillan Publishing Co., New York, NY, 1976.

Calkins, Carroll C. *Reader's Digest Illustrated Guide to Gardening*. The Reader's Digest Association, Inc. Pleasantville, NY.

Creasy, Rosalind. *The Complete Book of Edible Landscaping*. Sierra Club Books, San Francisco, CA. 1982.

Encyclopedia of Organic Gardening. Rodale Press, Inc., Emmaus, PA. 1978.

Getting the Bugs Out of Organic Gardening. Rodale Press, Inc., Emmaus, PA. 1973.

Halpin, Anne M., Ed. *Rodale's Home Gardening Library: Fruit*. Rodale Press, Emmaus, PA. 1988.

Jacob, Irene and Walter. *Gardens of North America and Hawaii: A Traveler's Guide*. Timber Press, Portland, OR.

Jagendorf, M. A. *Folk Wines, Cordials, and Brandies*. The Vanguard Press, Inc., New York, NY.

Jaynes, Richard, Ed. *Nut Tree Culture In North America*. Nothern Nut Growers Assoc., Hamden, CT. 1979.

Kourik, Robert. *Designing and Maintaining Your Edible Landscape Naturally*. Metamorphic Press, Santa Rosa, CA 95402. 1986.

Leeg, Elizabeth. *Container Farming: Grow Your Own Food Indoors, on Patios, Terraces, and in Small Yards*. Major Books, Chatsworth, CA 91311. 1975.

Logan, Harry Britton. *A Traveler's Guide to North American Gardens*. Charles Scribner's Sons, New York, NY. 1974.

Logan, William Bryant. *The Gardener's Book of Sources*. Viking Penguin, Inc. New York, NY. 1988.

MacCubbin, Tom. *Florida Home Grown*. Sentinel Communications, Inc. Orlando, FL. 1987.

Maxwell, Lewis S. and Betty M. *Florida Fruit*. Lewis S. Maxwell Publisher, Tampa, FL. 1967.

McEachern, George Ray. *Growing Fruits, Berries and Nuts in the South*. Gulf Publishing Co. Houston, TX. 1978.

Organic Gardeners Complete Guide to Vegetables & Fruit. Rodale Press, Emmaus, PA. 1982.

Roy, Richard and Lance Walheim. *Citrus: How to Select, Grow and Enjoy*. HP Books, Tucson, AZ. 1981.

Southwick, Lawrence. *Dwarf Fruit Trees For the Home Gardener*. Garden Way Publishing, Charlotte, VT. 1972.

Stebbins, Robert, and Michael MacCaskey. *Pruning: How-to Guide for Gardeners*. HP Books, Tucson, AZ. 1983.

Stebbins, Robert, and Lance Walheim. *Western Fruit, Berries and Nuts*. HP Books, Tucson, AZ. 1981.

The Sweet Apple Gardening Book. Doubleday & Co., Garden City, NY. 1972.

Tukey, Harold B. *Dwarfed Fruit Trees*. Macmillan Co., New York. 1964.

Van Atta, Marian. *Growing and Using Exotic Foods.* Pineapple Press, Inc., Sarasota, FL. 1991.

Van Atta, Marian. *Living Off the Land, Space Age Homesteading.* Pine & Palm Press, Melbourne, FL. 1973.

Watkins, John V., and Thomas J. Sheehan. *Florida Landscape Plants,* Rev. Ed. The University Presses of Florida, Gainesville, FL. 1965/1975.

Your Garden Homestead. Houghton Mifflin Co., Boston, MA. 1977.

CATALOG OF GARDENING BOOKS

Garden Way Publishing
Storey Communications, Inc.
Schoolhouse Road
Pownal, Vermont 05261

PERIODICALS *(Magazines, Journals, Reports, Newsletters)*

Applesource. Tom Vorbeck, Rt. 1, Chapin, IL 62628.

California Rare Fruit Growers Quarterly Newsletter. The Fullerton Arboretum, California State University, Fullerton, CA.

Friends of the Trees. P.O. Box 1466, Chelan, WA 98816.

Gil's Garden. 1300 Casa Vista Drive, Palm Harbor, Florida 34683.

Horticulture Magazine. 20 Park Plaza, Suite 212, Boston, MA 02116.

International Dwarf Fruit Tree Assoc. Newsletter. 301 Horticulture Bldg., Michigan State Univ., East Lansing, MI 48824.

International Tree Crops Institute USA, Inc. Newsletter. Route #1, Black Lick Rd., Gravel Switch, KY 40328.

Living Off the Land, Subtropic Newsletter. Marian Van Atta, Ed., Geraventure Corp., P.O. Box 2131, Melbourne, FL 32902-2131.

Living Tree Center Newsletter. P.O. Box 797, Bolinas, CA 94924.

National Gardening Assoc. Magazine. 180 Flynn Ave., Burlington, VT 05401.

New York State Fruit Testing Cooperative Assoc. Newsletter. Geneva, NY 14456.

Organic Gardening Magazine. 33 E. Minor St., Emmaus, PA 18049.

Rodale Research Center Reports. Box 323, RD#1, Kutztown, PA 19530.

Southern Living. Box 523, 2100 Lakeshore Drive, Birmingham, AL 35201.

Sunset Magazine. Sunset Books, Lane Publishing Co., Menlo Park, CA 94025.

Tropical Fruit News. Rare Fruit Council International, P.O. Box 561914, Miami, FL 33256.

Western Cascade Tree Fruit Association Newsletter. 9210 131 St. N.E., Lake Stevens, WA 98258.

Western Fruit Grower. Meister Publishing Co., 37841 Euclid Ave., Willoughby, OH 44094.

ABOUT THE AUTHORS

Marian Van Atta

Born in Milwaukee, Wisconsin, Marian Van Atta has lived and planted fruit trees in Wisconsin, California, Florida, British Columbia and North Carolina. Dubbed "Florida's Mother Nature" by the *Miami Herald*, she is nationally known as a home gardening advocate.

For 22 years, Van Atta has written the weekly "Living Off the Land" food and gardening column carried by newspapers around the United States and Canada. Her "Living Off the Land" newsletter has subscribers the world over. Van Atta has written articles for *Organic Gardening, National Garden,* and *Weekend Gardening,* as well as *Golden Years, Florida Gardener,* and *Brevard Senior Citizen & Home Gardener.* She is the author of three books: *Living Off The Land, Space Age Homesteading; Wild Edibles, ID for Living Off The Land;* and *Growing & Using Exotic Foods.*

On the lecture circuit, Van Atta presents gardening programs for all ages from preschool to high school, as well as for garden clubs, rare fruit growers, and retirement clubs. She often appears on radio and TV. Her aim is to make the growing of fruits and nuts a vital part of family living.

She is married to Jack Van Atta, fellow gardening enthusiast, and they have four grown children and four grandchildren.

Shirley L. Wagner

Shirley Wagner is a writing and editing consultant with more than 15 years experience in the field. She holds an M.A. Degree in English from the University of Florida and has taught writing and editing at the college level for a number of years. She has won prizes and recognition for her poetry, plays and articles on a variety of subjects and is editor of three association newsletters and past editor of a 52-page business magazine. She was president of the Cape Canaveral Branch of The National League of American Pen Women, Inc. from 1990 to 1992.

Shirley and her husband Ed Shores enjoy living in Florida where they start their days with fresh orange juice from homegrown oranges chilled on the tree.

Index

Bold numbers indicate illustrations.

Sprinklers, 32
Squirrels, 66
Staking, 29
Strybing Arboretum, 114, 116
Stylar rot, 78
Sulphur, 22, 34, 36, 40
Super trees, 89-91

Tangelos, 54
  varieties of, 80
Tangerines, varieties of, 79-80
Tansy, 41
Terrain, 19-20
Thrips, 40, 41
Timing,
  in fertilizing, 34-36
  in harvesting, 44
  in planting, 26-27
  in pruning, 38
Toads, 106
Trace elements, 34, 36
Trail mix, 52
Training (see Pruning)
Traps, no-kill, 42, 106
Tree nurseries, 62-63

Tropical Apple Pie, 98

Utah, University of, State
  Arboretum, 114, 117

Vegetarian Tzimmes, 97

Walnut, English, 17, 91
Walnuts, 51
  care of, 68
  harvesting, 68
  planting tips, 68
  recipes, 69, 99
  varieties of, 67-68
Wasps, trichogramma, 41
Watering, 27, 31-33
Whipped Chestnuts, 97
White flies, 40, 49, 76
Whitton, Gil, 16, 73

Yields, 43
Yogurt Pan Tart, 101

Zinc, 34, 36, 66
Zones, planting, 15

Notes

Notes